Topics in Mining, Metallurgy and Materials Engineering

"Topics in Mining, Metallurgy and Materials Engineering" welcomes manuscripts in these three main focus areas: Extractive Metallurgy/Mineral Technology; Manufacturing Processes, and Materials Science and Technology. Manuscripts should present scientific solutions for technological problems. The three focus areas have a vertically lined multidisciplinarity, starting from mineral assets, their extraction and processing, their transformation into materials useful for the society, and their interaction with the environment.

**** Indexed by Scopus (2020) ****

Wenwen Song

Nano-Engineering of High Strength Steels

Springer

Wenwen Song
Granularity of Structural Information
in Materials Engineering
University of Kassel
Kassel, Germany

ISSN 2364-3293 ISSN 2364-3307 (electronic)
Topics in Mining, Metallurgy and Materials Engineering
ISBN 978-3-031-42966-8 ISBN 978-3-031-42967-5 (eBook)
https://doi.org/10.1007/978-3-031-42967-5

This Springer imprint is published by the registered company Springer Nature Switzerland AG
The registered company address is: Gewerbestrasse 11, 6330 Cham, Switzerland

篤學奮進

徐匡迪

二〇一五年二月

The process of scientific discovery is, in effect, a continual flight from wonder.

Albert Einstein

To my beloved family and our sunny life

♡♡♡

Foreword

The medieval blacksmith takes the iron from the anvil, carefully examines the color in the dark surroundings of his forge, then hammers, heats, murmurs traditional timekeeping sayings, checks the color again to estimate the temperature, and then quenches the sword into water. He produces high strength steel with a nanostructure—even if he would never call it that. So, what's new about a book on nano-engineering of high strength steels?

The science is new. From the enigmatic, myth-shrouded process, a scientific field of work has emerged with new analysis devices, new observations, new process paths, new combinations of properties, and finally new theories. In the beginning there was the assumption that new insights into material behavior, for example, into the martensitic structure of the forged sword, could be found on a higher-resolution scale beyond classic light microscopy. The development of a new generation of analysis devices, examples of which are high-resolution electron microscopy and atom probe tomography, led to the fact that crystallographic defect structures and chemical heterogeneities could be described quantitatively with hitherto unknown resolution accuracy. New material insights were quickly derived from this, both for optimized combinations of properties and for new aspects of material behavior, e.g. the intensive use of stress- or strain-induced low-temperature phase transformations. The researchers benefited from the fortunate coincidence of various circumstances: the development of new scientific methods and the great demand for high strength steels for lightweight construction in the automotive industry.

And then, in the 1990s, the fascinating era of developing new high strength steels began for Materials Scientists. Everything just fitted together: the high-resolution investigation methods brought a multitude of new findings on the structure of materials. The demand in the automotive industry, the competition with specifically lighter materials, the availability of new plant technology—here the flexibility of the new continuous annealing process after cold rolling should be emphasized above all—ensured that new steel developments quickly found their way from laboratory to industry. In addition, there was the idea of giving these materials the new name Advanced High Strength Steels, soon to be supplemented by the term 1st, 2nd, or

3rd generation. Science, technology, and marketing entered into a happy symbiosis here.

After years of rapid growth, constantly new findings, many new theoretical approaches, now is the time to think, to classify, and to evaluate the observed phenomena and thus to lay a foundation for future research work but also for teaching. This is precisely the task of this book.

First, various aspects of nanostructuring are presented in an overview. Based on this, the special features of the Advanced High Strength Steels are described and the impact of nanostructuring on properties are explained. Important new methods for quantifying nanostructures are presented. Finally, in the chapters Precipitation Engineering, Interface Engineering, and Short-Range Ordering Engineering, procedures for the development of nanostructured materials are compiled.

The young scientist Wenwen Song completed a Master's degree in Materials Science in Engineering at Shanghai University and a parallel Bachelor's degree in Business Management at Shanghai University of Finance and Economics. She received a Dr.-Ing. degree of RWTH Aachen University. Her thesis has been prepared in cooperation with ICAMS (International Center for Advanced Materials Simulation at the University of Bochum). Ms. Song now heads the working group on nanostructured materials at the Steel Institute of RWTH Aachen University, oversees the institute's cooperation with federal research centers, and is responsible for the atom probe laboratory. Her scientific work is characterized by an exceptional breadth of methodological approaches to material development using among others high-resolution electron microscopy, synchrotron and neutron radiation and ab initio material modelling. Based on her own work and considering the current state of knowledge, in her habilitation thesis—the basis for this book publication—she describes the physical principles for a number of newly developed high-performance materials and outlines possible future applications.

The optimism of the discoverer floats through this book. Today's voyages of discovery in the discipline of Materials Science and Engineering do not go to new elements of the periodic table but to scales of structural and chemical resolution not accessible in the past. It is a fascinating new world that is opening up for materials researchers here. The book is a way to systematically explore and use the nm-scale observations and phenomena for present and future structural materials. It encourages discovery and will hopefully stimulate many young people to enter this new world of materials. Thus, I wish this book a wide circulation.

Wolfgang Bleck
Steel Institute
RWTH Aachen University
Aachen, Germany

Preface

Since many years, I've been deeply attracted by the fascinating features of the metallic materials in a nano-world. When approaching to a material, I'm very curious about how it would look like at different scales and what makes it to be "the" material. In my eyes, every material has its unique characteristics and unique properties to be developed with its full potential for various applications. High strength steels are a group of metallic materials that exhibit excellent properties for various industrial applications in automotive, aerospace, infrastructure, etc. And they are the important member of future materials to build up the future smart cities. This book is composed on the basis of my habilitation thesis study and the research work was carried out during my service as the group leader of Nanostructured Materials at the Steel Institute (IEHK) in RWTH Aachen University.

Above all, I would like to thank my Ph.D. supervisor, Prof. Wolfgang Bleck from the Steel Institute at RWTH Aachen Unviersity and my mentor, Prof. Kuangdi Xu from the Chinese Academy of Engineering. The treasurable word "笃学奋进" (permanently pursuing serious science) from Prof. Xu is of great encouragement for me when I started my postdoctoral research work in 2015 and has been a motto with me all along the way during my research. I deeply appreciate my Ph.D. supervisor, Prof. Wolfgang Bleck, for his permanent support and fruitful discussions within the research work in this book. Their great supports and suggestions inspired me with brilliant scientific ideas to initiate and carry out my research. The work would not appear as it is, without their encouraging comments and valuable remarks.

I am so grateful for the precious opportunity to do the research work at IEHK within the Collaborative Research Centre (SFB) 761—"Steel—Ab initio: Quantum mechanics guided the design of new Fe-based materials" with cooperation with Max-Planck-Institut für Eisenforschung GmbH (MPIE), Institute of Inorganic Chemistry at RWTH Aachen University (IAC, RWTH Aachen), Central Facility for Electron Microscopy at RWTH Aachen University (GFE, RWTH Aachen), Institute of Materials Chemistry at RWTH Aachen University (MCH, RWTH Aachen), Institute for physical metallurgy and materials physics at RWTH Aachen University (IMM, RWTH Aachen), Institute of metal forming at RWTH Aachen University (IBF,

RWTH Aachen), Institute for Materials Applications in Mechanical Engineering at RWTH Aachen University (IWM, RWTH Aachen), Forschungszentrum Jülich, ThyssenKrupp Steel Europe AG, Faurecia Autositze GmbH, ThyssenKrupp Kohenlimburg GmbH, Salzgitter Mannesmann Forschung GmbH. In particular, I would like to express the deep thanks for the continuous discussions and inspired exchange with Prof. Richard Dronskowski at IAC, Prof. Joachim Mayer at GFE and FZJ, Prof. Dierk Raabe at MPIE on the topic of nanostructuring and nano-engineering the advanced high strength steels, and Prof. Ulrich Krupp for his kind support and helpful comments on writing this book.

Furthermore, the experimental support from beamline P02.1 at DESY, Hamburg and the KWS-2 beamline of Jülich Centre for Neutron Science (JCNS) at Heinz Maier-Leibnitz Zentrum (MLZ) is gratefully acknowledged. And I would like to thank the financial support within the SFB761 "Steel—Ab initio" project and the "Nachwuchsakademie 2016—Thermodynamics and kinetics in multi-component metallic and ceramic materials" granted by the Deutsche Forschungsgemeinschaft (DFG).

I would like to express special thanks to the very inspiring discussion with Sir Prof. Harry Bhadeshia and Dr. Enrique Galindo-Nava from the University of Cambridge on the topic of bainite and martensite formation, Prof. Pedro Rivera-Diaz-del-Castillo from the University of Lancaster on the topic of bearing steels, Prof. George Smith and Prof. Michael Moody from the University of Oxford on the topic of short-range ordering analysis by atom probe tomography, Prof. Lei Lu from the Metals Institute of Chinese Academy of Science and Prof. Robert Ritchie from the University of California, Berkeley, on the topic of novel nanostructured metallic materials design.

I am very thankful to my colleagues at IEHK for their support and I enjoy the working atmosphere very much at RWTH Aachen University. I deeply appreciate the engineers and technicians at RWTH Aachen for their effort to accelerate a sound and substantial progress. My deep gratitude particularly goes to my students and group members, Wei Zhang, Hai Huang, Ahmet Bahadir Yidiz, Marc Ackermann, Lulu Liu, Alex Gramlich, Kuan Ding, Dilay Kibaroglu, Victor Nieto, Diego Naranjo, Asmaa Elbeltagy, Maria Mora-Acuna, Carsten Drouven, Yan Ma, Xiao Shen, Zigan Xu, Hannah Schwich, Jinxiong Hou, Fengqin Ji and my research assistants Oguz Gülbay, Bowen Zou and Markus Felten for their helpful daily supports and for sharing a common passion for scientific research work.

This work is dedicated to my Mum and Dad. They cherish me by all their love and support through my life. They taught me step by step since I was born and founded my desire to be great and enthusiastic for work and life. My deepest love goes to my husband Kai and my son Max Dechen. They locate my heart and soul and they light the central happiness of my life.

At last, I would like to sincerely appreciate this precious opportunity to work and live with the world-class scientists, colleagues, assistants, and friends during my scientific pursuit. I cherish all the people and experiences which make the last years excellent, wonderful, and unique in my life.

Aachen, Germany
July 2022

Wenwen Song

Contents

1 **Introduction** 1

2 **General Aspects of Nanostructure** 7
2.1 Nanostructures in Metallic Materials 7
2.1.1 Ultrafine-Grained Structure 7
2.1.2 Nano-Laminates 9
2.1.3 Nano-Plates 10
2.1.4 Nano-Precipitates 11
2.1.5 Nano-Twins 13
2.2 Processing of Nanostructured Metallic Materials 15
2.2.1 Severe Plastic Deformation (SPD) 15
2.2.2 Thermomechanical Treatment (TMT) Processing 15
2.2.3 Solid Reaction 17
2.3 Properties of Nanostructures 20
2.4 Applications of Nanostructures 22
2.5 Nano-Engineering of Metallic Materials 24
References 26

3 **High Strength Steels** 31
3.1 Development of High Strength Steels 31
3.1.1 1st Generation Advanced High Strength Steels 32
3.1.2 2nd Generation Advanced High Strength Steels 33
3.1.3 3rd Generation Advanced High Strength Steels 33
3.2 Deformation Mechanisms and Strain Hardening 34
3.2.1 Transformation-Induced Plasticity (TRIP) 34
3.2.2 Twinning-Induced Plasticity (TWIP) 36
3.2.3 Microband-Induced Plasticity (MBIP) 38
3.2.4 Dynamic Strain Aging (DSA) 40
3.3 SFE and 3D Deformation Mechanisms Map 41
3.3.1 Stacking Fault and Stacking Fault Energy 41
3.3.2 Experimental and Calculation Methods to Determine Stacking Fault Energy 41

3.3.3 Thermodynamic Description of the Steels in Fe-Mn-C and Fe-Mn-Al-C Systems ... 44
3.3.4 Thermodynamics-Based Deformation Mechanisms Maps Calculation ... 45
3.4 Prediction and Control of Strain Hardening Behaviour in High-Mn Austenitic Steels ... 47
3.4.1 Prediction of the Effect of Chemical Composition and Temperature on Mechanism Maps ... 47
3.4.2 Influence of Temperature on the Deformation Behaviour of Fe-0.3C-17Mn-1.5Al Steel ... 49
3.4.3 Influence of Chemical Composition on Deformation Behaviours of Fe-Mn-C and Fe-Mn-Al-C Steels with iso-SFE ... 52
3.4.4 Influence of Chemical Composition on Strain Hardening of Fe-Mn-C and Fe-Mn-Al-C Steels with iso-SFE ... 53
3.5 Summary ... 55
References ... 56

4 Nanostructure Characterization Methods ... 61
4.1 High Energy Synchrotron X-Ray Diffraction (SYXRD) ... 61
4.1.1 Principle of SYXRD ... 61
4.1.2 Instrumental Set-Up and Sample Preparation ... 63
4.1.3 Data Analysis Based on Rietveld Method ... 64
4.2 Small Angle Neutron Scattering (SANS) ... 66
4.2.1 Introduction to SANS ... 66
4.2.2 Experimental Set-Up and Sample Preparation ... 69
4.2.3 Data Analysis ... 70
4.3 Atom Probe Tomography (APT) ... 71
4.3.1 Development of APT ... 71
4.3.2 Principle of APT ... 72
4.3.3 Sample Preparation ... 74
4.3.4 3D Atoms Map Reconstruction ... 76
4.4 High-Resolution Transmission Electron Microscopy (HRTEM) ... 79
4.4.1 Introduction to Electron Crystallography ... 79
4.4.2 Principles of TEM/STEM ... 79
4.4.3 Electron Microscopy and Electron Diffraction ... 80
4.4.4 Sample Preparation and Measurements of HRTEM ... 83
4.5 Summary ... 84
References ... 85

5 Precipitation Engineering ... 89
5.1 Definition and History ... 89
5.2 Precipitation Strengthening Mechanisms ... 90
5.2.1 Precipitation Strengthening Mechanisms ... 90

5.2.2 Influences of Precipitate Characteristics on Strengthening 94
5.3 Nano-Precipitation Strengthening in Steels 99
5.4 Applications 102
5.4.1 Coherent Precipitation Strengthening in HMnS Steels 102
5.4.2 Microalloyed Carbides Strengthening in MMnS Steels 108
5.5 Summary 112
References 114

6 Interface Engineering 119
6.1 Definition and History 119
6.2 Interfaces Characteristics During Phase Transformations 120
6.3 Interface Engineering in High Strength Steels 121
6.4 Application 131
6.4.1 C and Mn Partitioning Across Bcc/Fcc Interface in MMnS Steels 131
6.4.2 Influence of Microstructural Morphology on Hydrogen Embrittlement 133
6.4.3 Carbon Enrichment at Nano-Twin Boundaries in Bearing Steels 140
6.5 Summary 142
References 144

7 Short-Range Ordering Engineering 147
7.1 Definition and History 147
7.2 Short-Range Ordering (SRO) in Metallic Materials 148
7.2.1 Formation of SRO in Metallic Materials 148
7.2.2 Microstructure Characterization of SRO in Metallic Materials 150
7.2.3 Influence of SRO on the Mechanical Properties in Metallic Materials 152
7.3 Applications 153
7.3.1 Study on a High Entropy Alloy by High Energy Synchrotron X-Ray Diffraction and SANS 153
7.3.2 Experimental Observation of Kappa Phase Formation Sequences by In-Situ Synchrotron Diffraction 157
7.3.3 Local Deformation and Mn-C Short-Range Ordering in a High-Mn Fe-18Mn-0.6C Steel 161
7.4 Summary 168
References 168

8 Conclusions and Final Remark 175

Abbreviations

2D	Two dimensional
3D	Three dimensional
AFM	Atomic force microscopy
AHSS	Advanced high strength steel
AIDE	Absorption-induced dislocation emission
APB	Antiphase boundary
APFIM	Atom probe field ion microscope
APT	Atom probe tomography
ART	Austenite reverted transformation
ATP	Advanced thermomechanical processing
AUS	Austenitization
bcc	Body-centred cubic
BF	Bright field
CALPHAD	Calculation of phase diagrams
CCT	Continuous cooling transformation
CN	Coordination numbers
CP	Complex phase
CRA	Cold-rolled-annealed
CRSS	Critical resolved shear stress
D&P	Deformed and partitioned
DESY	Deutsches Electron-Synchrotron centre
DF	Dark field
DFT	Density function theory
DIC	Digital image correlation
DIFT	Deformation-induced ferrite transformation
DP	Dual phase
DRV	Dynamic recovery
DRX	Dynamic recrystallization
DSA	Dynamic strain aging
EBSD	Electron backscatter diffraction
ECAP	Equal channel angular pressing

ECCI	Electron channelling contrast imaging
ED	Electron diffraction
EDS	Energy dispersive X-ray spectrum
EELS	Electron energy loss spectroscopy
EXAFS	Extended X-ray absorption fine structure
fcc	Face-centred cubic
FEM	Field electron microscopy
FIB	Focus ion beam
FIM	Field ion microscopy
FOV	Field of view
GB	Grain boundary
GND	Geometrically necessary dislocations
GOS	Grain orientation scattering
HAGB	High angle grain boundary
hcp	Hexagonal-close packed
HE	Hydrogen embrittlement
HEA	High entropy alloy
HEDE	Hydrogen-enhanced decohesion
HELP	Hydrogen enhanced localized plasticity
HMnS	High-Mn steels
HPT	High-pressure torsion
HRA	Hot-rolled-annealed
HRSTEM	High-resolution scanning transmission electron microscopy
HRTEM	High-resolution transmission electron microscopy
HSLA	High strength low alloy steel
IA	Intercritical annealing
IPF	Inverse pole figure
IQ	Image quality
LE	Local equilibrium
LEAP	Local electrode atom probe
LOM	Light optical microscope
LRFT	Long-fibre reinforced thermoplastic
LRO	Long-range ordering
MAUD	Materials analysis using diffraction
MBIP	Microband-induced plasticity
MC	Monte Carlo
MEA	Medium entropy alloy
M_f	Martensite finish temperature
MLZ	Heinz Maier-Leibnitz
MMnS	Medium-Mn steels
M_s	Martensite start temperature
NBD	Nano-beam electron diffraction
NN	Nearest-neighboured
NPLE	None (negligible) partitioning local equilibrium mode
NT	Nano-twin

PAGB	Prior austenite grain boundary
PDF	Pair distribution function
PE	Para-equilibrium
PLC	Portevin-Le Chatelier
PLE	Local equilibrium with partitioning
Q&P	Quenching and partitioning
ROI	Region of interests
SAED	Selected area electron diffraction
SANS	Small angle neutron scattering
SDD	Sample to detector distance
SE	Segregation engineering
SEM	Scanning electron microscope
SF	Stacking fault
SFE	Stacking fault energy
SGTE	Scientific Group Thermodata Europe
SLIP	Slipband-induced plasticity
SMAT	Surface mechanical attrition treatment
SPD	Severe plastic deformation
SRC	Short-range clusters
SRO	Short-range ordering
SSRT	Slow strain rate tensile
STEM	Scanning transmission electron microscopy
SUTS	Specific ultimate tensile strength
SYS	Specific yield stress
SYXRD	Synchrotron X-ray diffraction
TB	Twin boundary
TE	Torsion extrusion
TEM	Transmission electron microscopy
TKP	Transmission Kikuchi patterns
TMP	Thermo-mechanical processing
TMT	Thermomechanical treatment
TOF	Time-of-flight
ToFMS	Time-of-flight mass spectrometer
TRIP	Transformation-induced plasticity
TTT	Time-temperature-transformation
TWIP	Twinning-induced plasticity
UE	Uniform elongation
UFG	Ultrafine grained
USFE	Unstable stacking fault energies
UTS	Ultimate tensile strength
XAFS	X-ray absorption fine structure
XANES	X-ray absorption near edge structure
XCCA	X-ray cross correlation analysis
YS	Yield stress

Chapter 1
Introduction

The development of human society over the last 2000 years is closely related to the progressive development of new metallic materials. In the ancient period, the application and availability of metallic products are limited due to the lack of metallurgy and fabrication technologies. Nowadays, metallic materials with superior properties are applied intensively in every aspect in our life, e.g. in transportation products, constructions, communication devices, energy transport, household appliances, medical instruments, making it one of the most important materials in a sustainable human society.[1] The global challenges, e.g. green-house development, sustainable infrastructure and clean energy, intelligent production, always request the solutions from the interdisciplinary work and the innovative approaches.

Iron-based alloys, namely steels, have been one of the most important metallic materials for human beings. A specialty of iron and iron-based alloys is the polymorphism behaviour. In other words, the iron and iron-based alloys can transform into different crystallographic structures (e.g. body-centred cubic, face-centred cubic and hexagonal-close packed) in the solid-state during heat treatment or deformation. The crystallographic structures of pure iron can be differentiated into α-iron (body-centred cubic, below 911 °C), γ-iron (face-centred cubic, 911 °C–1392 °C), and δ-iron (body-centred cubic, 1392 °C–1535 °C) as a function of temperature (Fig. 1.1). The diversities in crystalline structure lead to the distinguished deformation behaviour, elastic properties, magnetic properties, electric properties etc. The Fe-C phase diagram is the most basic tool for steel design.

The kinetics of the transformation of supercooled austenite can be described using the time–temperature-transformation (TTT) diagram and continuous cooling transformation (CCT) diagram. The transformation temperature of pure iron is changed through the addition of alloying elements. The phase diagram provides a theoretical guideline of thermodynamics equilibrium conditions, which supports us to further

[1] J.M. Allwood, J.M. Cullen. Sustainable Materials: With Both Eyes Open. Uit Cambridge Ltd (2011).

W. Song, *Nano-Engineering of High Strength Steels*,
Topics in Mining, Metallurgy and Materials Engineering,
https://doi.org/10.1007/978-3-031-42967-5_1

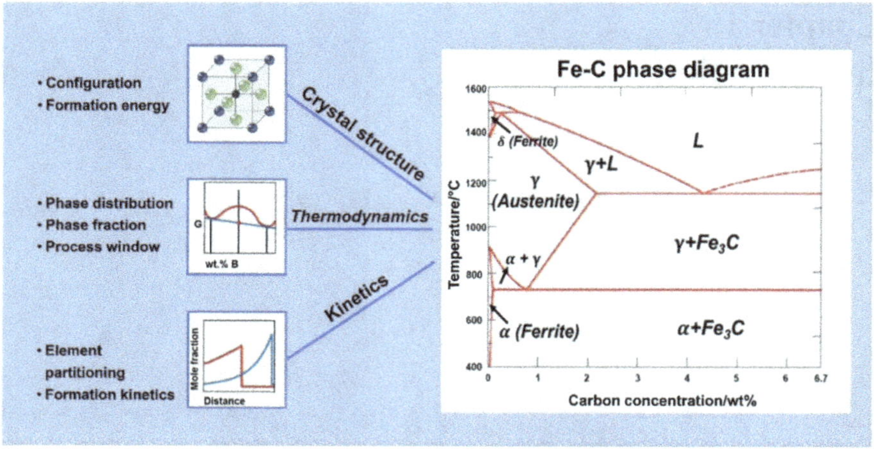

Fig. 1.1 Fe-C phase diagram, crystal structure, thermodynamics and kinetics

explore the phase transformation kinetics and microstructure evolution of alloys with different chemical compositions at a range of temperatures.

In the research of metallic materials, the linkage of structure-processing-property is considered as the very important principle to understand the alloys. Following this basic principle, one can further design, select and assess suitable materials for a specific application. Over the last decades, based on multi-scale understanding of the metallic materials—from metre (components) to micrometre (grains and phases) and further down to nanometre (second phases and stacking faults), a number of extraordinary metallic materials have been successfully developed and commonly applied, e.g. ultra-strong steels for automotive and aerospace applications, orthopaedic medical materials, etc. New material design concepts speed up the development of novel alloys with superior properties. With the recent development and application of the theoretical simulation tools and the advanced characterization methods, it has been found that it is necessary to understand the underlying mechanisms of phase transformations and deformation behaviours of the metallic materials at multi-scale in a correlative way (Fig. 1.2). So that one could figure out the key controlling parameters and to tune the materials properties.

The nano-engineering concept offers new opportunities to design and engineer the metallic materials into nano-sized structures with tailored properties. In recent years, several nano-engineering concepts have been proposed, e.g. segregation engineering, interface engineering, nano-precipitation engineering, etc. Segregation engineering is an effective approach to accelerate the phase transformation via targeted segregation at nanometer scale to defects in nanostructured metals. Interface engineering provides great opportunities to enhance mechanical properties by structural refinement via introducing multiple interfaces. Nano-precipitation engineering shows great potential to enhance the mechanical properties of metals by precipitation hardening effect.

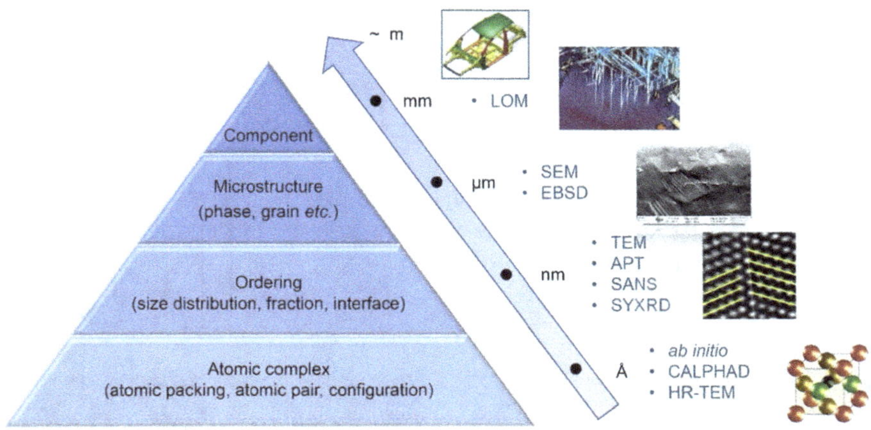

Fig. 1.2 Multi-scale approach in the development of metallic materials

So far, very small grain sizes can be obtained by thermo-mechanical processing (TMP) that combines deformation and transformation processes.[2] Ultrafine-grained structures with typical grain sizes of 1 μm or less can be obtained especially by severe plastic deformation. Different options for the development of nanostructured steels rely on various deformation mechanisms that continuously refine the microstructure during cold forming[2]. By the nano-engineering concept, the steels with a metastable austenite starting microstructure can be systematically refined during deformation and develop extraordinary service properties. Nano-structuring offers new chances for the design of material properties by diffusion and partitioning phenomena of interstitial and substitutional elements as well as by the interaction of dislocations with a great variety of interfaces[2]. The slipband-induced plasticity (SLIP) refers to the dislocation generation and reactions in the materials, which play a very important role in most of the metallic materials. The transformation-induced plasticity (TRIP) effect depends on the strain-induced austenite → martensite transformation[2]. The twinning-induced plasticity (TWIP) effect is a consequence of the introduction of nanometre-scale deformations induced twins. In microband-induced plasticity (MBIP) steels planar glide of dislocations occurs due to the formation of intermetallic phases, as shown in Fig. 1.3.

The central goal of this book is to provide deep insights of how to adjust the nanostructures in high strength steels in order to achieve enhanced mechanical properties. This work provides summarized state-of-the-art knowledge of the nano-engineering approaches, e.g. precipitation engineering, interface engineering, short-range ordering engineering. The nanostructure-property relationship in a series of high strength steels, e.g. TRIP/TWIP/MBIP in high-Mn steels (HMnS), medium-Mn steels (MMnS), bearing steels, tool steels, etc., were investigated. New methods that aid controlling the process of phase transformation during deformation and/

[2] W. Song, T. Ingendahl, W. Bleck, Control of strain hardening behavior in high-Mn austenitic steels, *Acta Metall. Sin. (Engl. Lett.)* 27 (2014) 546-556.

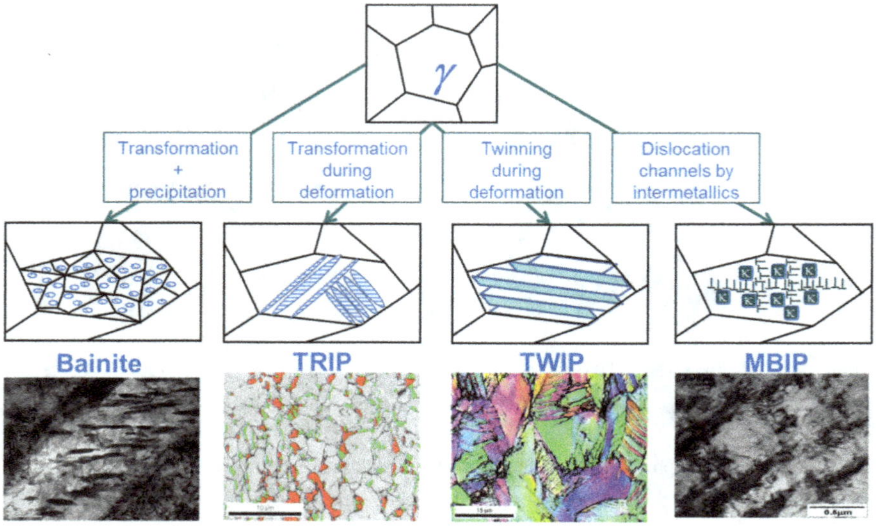

Fig. 1.3 Nano-engineered steels[2]

or thermal treatment in the steels were focused to be explored. In order to achieve the main goal of this thesis, a combination of experimental and simulation tools was employed, e.g. atom probe tomography (APT), high-resolution transmission electron microscopy (HRTEM), synchrotron and neutron radiations, ab initio calculations and Calphad thermodynamics simulations.

In this book, Chap. 2 introduces the general aspects of nanostructures, including the classification and characteristics of different type of nanostructures in metallic materials, the processing and properties of the nanostructures and an overview of the applications of nanostructures. Chapter 3 starts with a review of the development of high strength steels and introduces the basic deformation mechanisms of high strength steels, *i.e.* TRIP/TWIP effect, MBIP effect, dynamic strain aging (DSA). Chapter 3.3 reports own recent work on novel high strength steels design based on 3D deformation mechanisms maps. The prediction and control of strain hardening behaviours in high strength steels were introduced in Chap. 3.4. Novel nanostructure characterization methods, *i.e.* high energy synchrotron X-ray diffraction (SYXRD), small angle neutron scattering (SANS), APT and HRTEM, were further described in Chap. 4, including the principles, instrumental set-up, sample preparation and data analysis. In Chaps. 5 and 6, the aforementioned precipitation engineering and the interface engineering approaches were introduced, respectively. Chapter 5 starts with the definition and history of the precipitation strengthening mechanisms, followed by research examples of nano-precipitation strengthening in steels. Own-related research work of precipitation engineering is then reported as application examples in Chap. 5.4. Chapter 6 introduces the interface engineering approach, starting with the definition and history of interface engineering. Grain boundary engineering approach and segregation engineering approach are not separated from this chapter.

Own-related research work is described as application examples in Chap. 6.4. These aforementioned novel concepts stimulate and inspired me to research the nanostructured metals and look for new strengthening concepts to improve mechanical properties. Recently, the successful attempt to employ short-range ordering (SRO) concept in various steels allows me to show a new pathway to overcome the strength-ductility trade-off in nanostructured metallic materials, which was awarded "Steel Innovation Prize 2018" in Berlin, Germany. This engineering concept is introduced in Chap. 7.

On the one hand, this book summaries the state-of-the-art knowledge of the nano-engineering approaches in literature, e.g. precipitation engineering approach, interface engineering approach, short-range ordering engineering approach. On the other hand, it includes my research contribution with nano-engineering case studies introduced in Chaps. 2–7. Own publication-related research contribution topics include:

- SFE and 3D deformation mechanisms map (Chap. 3.3)
- Prediction and control of strain hardening behaviour in high-Mn austenitic steels (Chap. 3.4)
- Coherent precipitation strengthening in HMnS steels (Chap. 5.4.1)
- Microalloyed carbides strengthening in MMnS steels (Chap. 5.4.2)
- C and Mn partitioning across bcc/fcc interface in MMnS steels (Chap. 6.4.1)
- Influence of microstructural morphology on hydrogen embrittlement (Chap. 6.4.2)
- Carbon enrichment at nano-twin boundaries in bearing steels (Chap. 6.4.3)
- Study on a high entropy alloy by high energy synchrotron X-ray diffraction and SANS (Chap. 7.3.1)
- Experimental observation of kappa phase formation sequences by *in-situ* synchrotron diffraction (Chap. 7.3.2)
- Local deformation and Mn-C short-range ordering in a high-Mn Fe-18Mn-0.6C steel (Chap. 7.3.3)

Chapter 2
General Aspects of Nanostructure

The state-of-the-art development of materials with high performance requests multi-scale understanding of the materials structures. Nanostructured materials exhibit unique characteristics and properties other than quantum materials and bulk material, Fig. 2.1. In macroscopic scale, the distinguished properties of nanostructured materials include high-performance mechanical properties, various functional properties (magnetic properties, optics properties, electrical properties etc.) and chemical properties. This chapter provides a detailed description on the general aspects of nanostructures, including the classification and characteristics of different types of nanostructures, the processing and properties of the nanostructures, as well as an overview of the applications of nanostructures.

2.1 Nanostructures in Metallic Materials

2.1.1 Ultrafine-Grained Structure

Grain refinement is one of the most effective methods which may improve both strength and toughness simultaneously, among the strengthening mechanisms in alloys. Ultrafine-grained steels with lean chemical compositions, strengthened primarily by grain refinement, have great potential for replacing some of the conventional low alloyed high strength steels. It may avoid or decrease the needs of additional alloying and/or heat treatments and to improve weldability. Figure 2.2 shows the typical ultrafine-grained microstructure in a Fe-0.2C-10.2Mn-2.8Al-1.5Si medium-Mn steel. An austenite plus ferrite duplex microstructure with globular grains is formed after intercritical annealing (Fig. 2.2a–e) [2]. The microstructure is almost fully recrystallized with only a few stacking faults (SFs) and dislocations, as revealed by electron channelling contrast imaging (ECCI) (Fig. 2.2f) technique [2]. ECCI

W. Song, *Nano-Engineering of High Strength Steels*,
Topics in Mining, Metallurgy and Materials Engineering,
https://doi.org/10.1007/978-3-031-42967-5_2

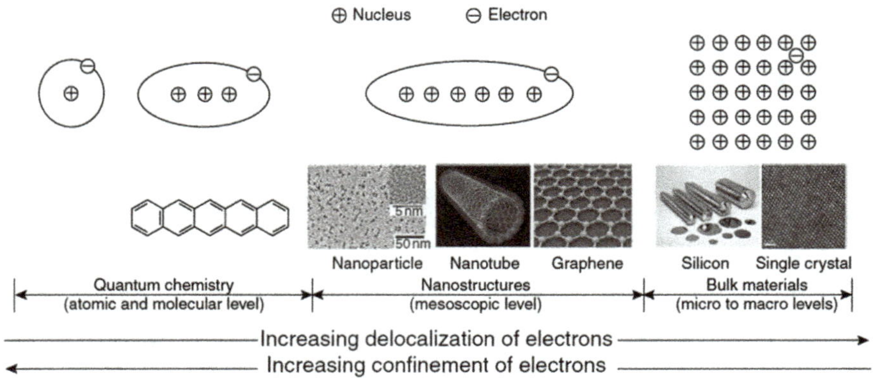

Fig. 2.1 Schematic of electron confinement from atom to bulk materials [1]

technique is an imaging method in SEM based on electron channelling effect, which is used for direct observation of lattice defects. ECCI shows comparable contrast features to dark-field transmission electron microscopy (TEM) with the advantages of large sample scale. The average density of dislocations is estimated to be around 5.7×10^{12} m^{-2} and 5.9×10^{12} m^{-2} for the samples annealed for 3 min and 60 min, respectively, based on ECCI observations of more than 50 individual grains for each sample [2].

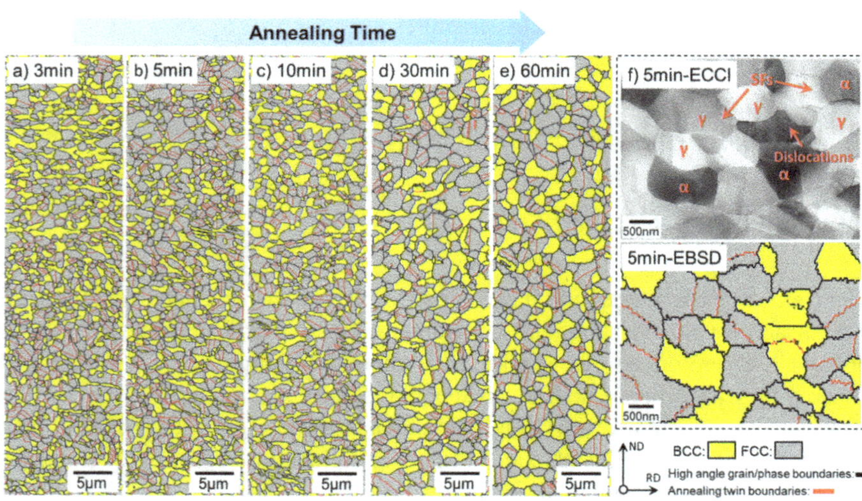

Fig. 2.2 EBSD phase mapping of the steel samples intercritically annealed at 800 °C for **a** 3 min, **b** 5 min, **c** 10 min, **d** 30 min and **e** 60 min; **f** Correlative ECCI and EBSD results for the 5 min annealed sample, showing a fully recrystallized microstructure with a few stacking faults (SFs) inside austenite and dislocations in ferrite [2]

2.1.2 Nano-Laminates

Nano-scale laminates occur in various steels, in particular in the pearlitic microstructure. During heavy deformation, the laminate microstructure is largely refined down to nanometre scale and provide ultra-high strength of the materials. For example, after heavy cold-drawing, the pearlitic steels can exhibit tensile strength higher than 5 GPa. Figure 2.3 shows the atom probe tomographic characterization of a hypereutectoid steel (Fe-0.98 C-0.31Mn-0.20Si-0.20Cr-0.01Cu-0.006P-0.007S in wt.%) with increasing true (logarithmic) strains. Carbon-enriched regions identified by green iso-concentration surfaces represent cementite, with the carbon-depleted regions being ferrite [3]. With increasing the drawing strain, firstly the volume fraction of cementite continuously decreases due to its mechanically driven chemical decomposition and secondly the carbon atoms are released from the dissolving cementite and are mechanically alloyed into the ferrite [3]. This leads to a deformation-driven carbon supersaturation of the ferrite. Third, the initially two-phase lamellar pearlite structure evolves, due to its dissolved cementite layers, into a carbon-decorated ferrite subgrain structure, as is visualized by the carbon segregation at the ferrite boundaries (marked by blue arrows) [3].

Upon straining, the pearlitic lamellar structure is refined down to nanometre scale with a concurrent decomposition of cementite. The cementite decomposition is reported to increase the average carbon concentration in ferrite far above the equilibrium concentration and to provide an additional strengthening mechanism due to

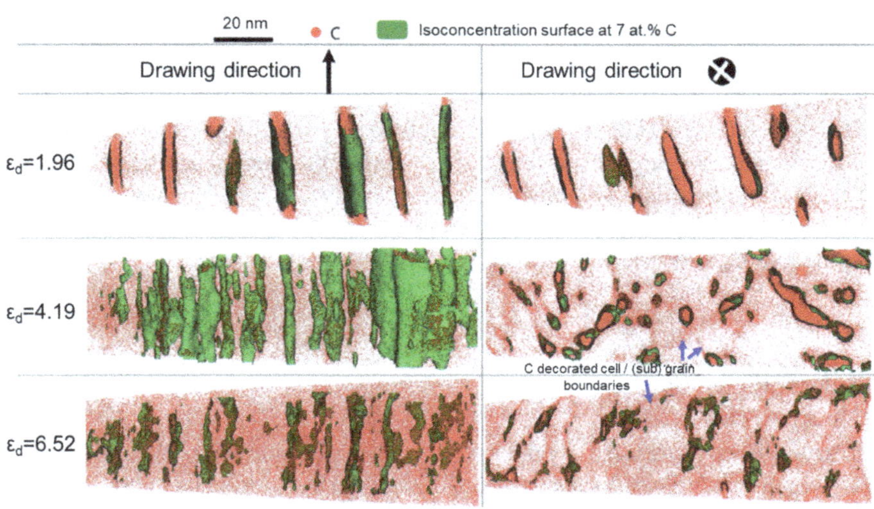

Fig. 2.3 Atom probe tomographic characterization of pearlitic steel wires cold drawn to different drawing strains. 3D carbon atom maps in both longitudinal (parallel to the drawing direction) and transverse cross section views (perpendicular to the drawing direction). Blue arrows mark some of the subgrain boundaries decorated with carbon atoms [3]

mechanical alloying [4]. Hot-rolled laminated steels may exhibit a good combination of mechanical properties [5]. Correlative work employing high energy beamline revealed that the strain partitioning between austenite and martensite was governed by a highly dynamical interplay of dislocation slip, deformation-induced phase transformation (*i.e.* causing the transformation-induced plasticity (TRIP) effect) and mechanical twinning (*i.e.* causing the twinning-induced plasticity effect). The nano-laminate microstructure morphology leads to enhanced damage resistance. Both effects result in enhanced strain-hardening capacity and damage resistance, and therefore, improved ductility [6].

The metallic crystalline-amorphous nano-laminate composites are considered to be both ductile and strong. Guo et al. synthesized and indented nano-laminate material consisting of nano-crystalline Cu and amorphous $Cu_{54}Zr_{46}$. The thickness of the crystalline Cu and amorphous CuZr layers are 10 and 100 nm, respectively [7]. After Vickers indentation, the CuZr/Cu multilayer composite was subjected to the correlative analysis by TEM and atom probe tomography (APT) to study nanostructural and atomic-scale chemical deformation effects, in which sharp shear bands around the pileup regions are observed in Fig. 2.4a–c [7]. The Cu layers are not significantly reduced in thickness in most pileup regions. Figure 2.4d, e shows the correlative TEM and APT analysis. The area in the bright field TEM micrograph of an APT specimen highlighted in yellow is the volume containing a shear band analysed by APT, Fig. 2.4d [7]. The yellow lines mark a region representing the volume investigated by APT, although the actual APT data in Fig. 2.4e are taken from a slightly different region in the same sample [7]. The white rectangles in Fig. 2.4d indicate regions from which nano-beam electron diffraction (NBD) patterns are acquired, which show crystalline diffraction spots in the Cu region and an amorphous halo in the CuZr region [7]. Figure 2.4e reveals the atomic composition of the sheared region taken by APT together with the crystallographic orientation obtained from TEM [7]. It shows that the {111} planes of the Cu are parallel to the interface plane of the shear band. The large displacement (>30 nm) of the sheared and fragmented Cu layer shown in Fig. 2.4e suggests that large numbers of Cu dislocations accommodate the shear inside the crystalline phase, extending it across the Cu/CuZr interface into the initially amorphous layer [7].

2.1.3 Nano-Plates

The nano-plates in steels mainly refer to the nano martensitic and bainitic plates. The bcc structure undergoes displacive transformation and is formed as ultrafine plate-like structures to minimize the large elastic strain energy. Development and upscaling of nano-grained bainitic steels show an outstanding wear resistance and competitive values for fatigue strength. The retained austenite may transform under deformation to martensite as TEM images show in Fig. 2.5 [8]. The Fe-1.0C-2.5Si-1.0Mn steel exhibits bainitic microstructure after bainitic isothermal treatment at 220 °C for 22 h [8]. The mechanical stability of austenite, *i.e.*, its ability to remain untransformed

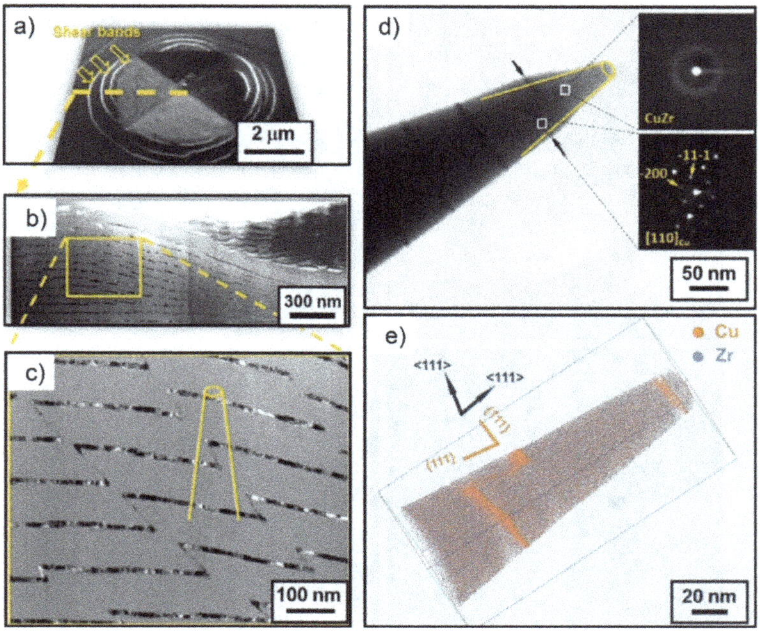

Fig. 2.4 **a** SEM view of CuZr/Cu multilayers after Vickers indentation with a load of 10 g; **b** Cross-sectional view of shear bands beneath an indent; **c** Bright field TEM image of an area containing shear bands; The yellow lines mark a region investigated by APT; **d** Bright field TEM image of an APT tip with corresponding CuZr and Cu nano-beam diffraction patterns. The arrows point to the shear band region; **e** APT reconstructed volume of APT tip [7]

during deformation, is mainly governed by its local chemical composition, its shape and size, crystal texture and the local constraint of the surrounding ferritic matrix [8].

2.1.4 Nano-Precipitates

Nano-precipitates are often incorporated to strengthen the alloys without sacrificing ductility. Jiang et al. reported the precipitation strengthening by Ni (Al, Fe) intermetallics in an aged steel Fe-18Ni-3Al-4Mo-0.8Nb-0.08C-0.01B (wt.%) [9]. A ultimate tensile strength (UTS) of up to 2.2 GPa and elongation of 8.2% are achieved after aging at 500 °C for 3 h. The outstanding strength is contributed largely from the semi-coherent precipitates. Details of the microstructure and elemental composition of the precipitates were characterized by APT, as shown in Fig. 2.6 [9]. Figure 2.6a shows the tomographic reconstruction from one of the APT datasets, revealing a large volume fraction of precipitates with a number density of about 3.7×10^{24} m^{-3}, as highlighted by an iso-concentration surface. The number density of nano-precipitates

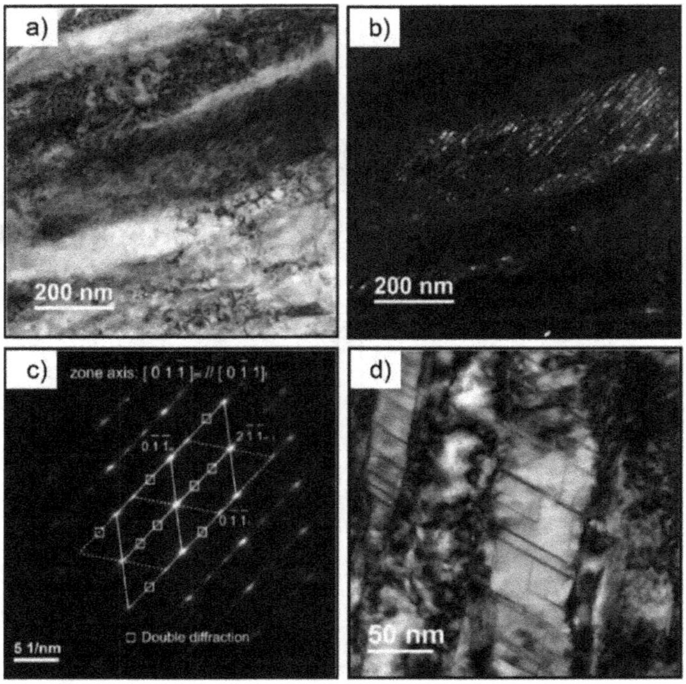

Fig. 2.5 **a** Bright; and **b** Dark field TEM images; and **c** Corresponding diffraction pattern, of twinned martensite in bainite formed at 220 °C for 22 h and deformed up to 2% strain in a high-C (1.0 wt.%) and high-silicon (2.5 wt.%) steel; **d** Bright field micrograph of mechanical twins in austenite in the same bainitic structure deformed up to 6% strain [8]

in the alloy is several times higher than that reported for conventional precipitation-hardened metallic materials [9]. This observation shows that lattice misfit design enables the formation of nano-precipitates with an extremely high number density. Although most precipitates assume a near-spherical shape, few precipitates are elongated. Several linear features can be revealed by the iso-concentration envelopes surrounding zones in the data containing a high concentration of Mo and interstitial impurities (C, B and P), as shown in Fig. 2.6b [9].

The larger precipitates imaged are mostly found in the vicinity of, or in contact with these dislocations, and tend to be more elongated (Fig. 2.6c). By making use of proximity histograms, calculated from the iso-concentration surfaces shown in Fig. 2.6d, the compositions of precipitates with various shapes are obtained [9]. The chemical composition of the precipitates enables a substantial reduction in cost compared to conventional maraging steels owing to the replacement of the essential but high-cost alloying elements cobalt and titanium with inexpensive and lightweight aluminium [9]. Strengthening of this class of steels is based on minimal lattice misfit to achieve

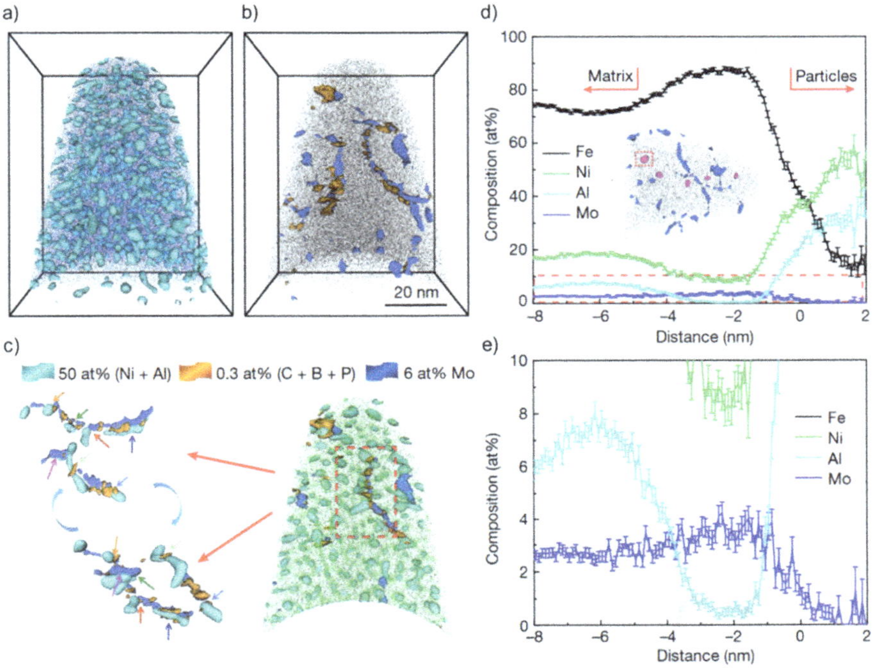

Fig. 2.6 APT analysis of the precipitates: **a** Precipitates highlighted by an iso-concentration surface (Ni + Al) = 50 at.%; **b** Regions with the iso-concentration surfaces (C + B + P) = 0.3 at.% and Mo = 6 at.% highlighting the presence of segregation zones along dislocations; **c** Precipitates growing near or at dislocations; **d** Proxigram showing the concentration profile across the selected precipitate; and **e** corresponding zoom-in image [9]

maximal precipitate dispersion and high cutting stress (the stress required for dislocations to cut through coherent precipitates and thus produce plastic deformation) [9].

2.1.5 Nano-Twins

Grain refinement to a nano-size will increase the number of grain boundaries, which can impede the dislocation motion during plastic deformation [10]. Such behaviour plays a significant role in achieving ultra-high strength. Intensive efforts have been made to explore strategies for simultaneously improving the strength and ductility in nano-crystalline metals by introducing nano-twin structures. When such nano-twin structures are introduced into the crystalline metals, which means that a high density of twin boundaries is embedded in the grains, acting as coherent boundaries to improve the mechanical and physical properties [11]. The twinning-induced plasticity (TWIP) effect enables designing austenitic Fe-Mn-C based steels with >70%

elongation at an ultimate tensile strength of >1 GPa. These alloys are characterized by high strain hardening coefficients owing to the formation of twins and complex dislocation substructures which dynamically reduce the dislocation mean free path. Further insight in the strain hardening mechanisms can be gained by a conventional TEM analysis of highly strained samples of a stainless Fe-Cr-Mn-C-N steel [12].

The primary mechanical twins contribute to the strain-hardening by acting as obstacles to dislocations gliding on different slip systems, which shows mechanical twins and intersecting stacking faults after 0.21 logarithmic strain. Dislocation accumulation at twin boundaries is observed [12]. Another case of a deformation-twinned nano-crystalline Pd grain which is shown in Fig. 2.7 [13]. Thin twin lamellae were formed during cold rolling. The two boxes in Fig. 2.7a indicate the areas for detailed defect analyses. Dislocations located in the grain interior (matrix and twin) are seen in Fig. 2.7b [13]. Figure 2.7c shows the highly strained zone along the twin boundary. Fourier transform was employed to calibrate the g-vectors and to calculate the strain tensor components of the matrix, twin respectively [13].

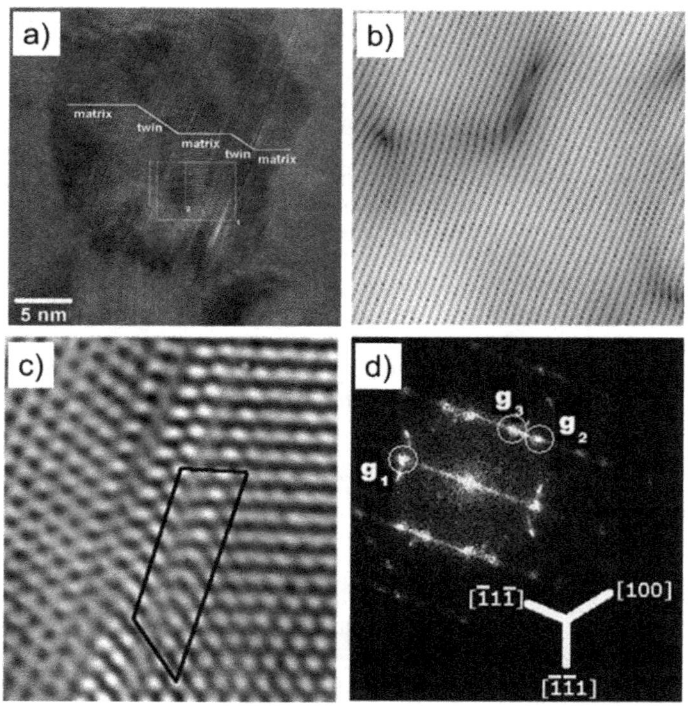

Fig. 2.7 **a** Lattice image of a deformation-twinned nano-crystalline Pd grain projected along [011], **b** Detail of **a** marked by box 1, **c** Detail of **a** marked by box 2, **d** Fourier transform of the former lattice image indicating the reflection of the [011] zone axis [13]

2.2 Processing of Nanostructured Metallic Materials

2.2.1 *Severe Plastic Deformation (SPD)*

In 1950s, P. W. Bridgman introduced a new technique of unlimited straining by torsion and compression to form dramatic grain refined structures, which is later defined as severe plastic deformation (SPD) method [14]. SPD method has been applied to produce nanostructure in metallic materials in the last decades, involving very large plastic strains typically in a complex stress state or high shear state [15, 16]. Nutting [17] firstly applied the microstructural analysis on SPD processed specimen and described the microstructure as sub-micron, near-equiaxed and dislocation-free grains with sharp high-angle boundaries. Further research confirmed this observation [18–22]. Especially the significant increase of yield stress and ultimate tensile strength in combination with sufficiently good ductility is demonstrated in the equal channel angular pressing (ECAP) processed materials [23, 24]. According to the different processing configurations, SPD methods include different processing technique, *e.g.* the high-pressure torsion (HPT), multidirectional forging (MDF), ECAP, accumulated roll bonding, torsion extrusion (TE). Each configuration has their typical technical features and processing parameters. For conventional metallic materials the strength might rise to a factor of 8 with up to 30–50% ductility [15, 25]. This strength enhancement is caused by the refined grain size in the submicrometre and nanometre ranges [26]. Figure 2.8 shows representative results obtained using orientation imaging microscopy and electron backscatter diffraction on high purity (99.99%) aluminium after processing by ECAP for up to a maximum of 12 passes [27]. After the first pass in Fig. 2.8b, the elongated subgrains are distributed in an array. Meanwhile, the microstructure gradually forms into the reasonably equiaxed grains separated by boundaries with high angles of misorientation [27]. The average grain size is reported to be ~1.2 μm after 12 passes and the fraction of high-angle boundaries is estimated at 74% [27].

2.2.2 *Thermomechanical Treatment (TMT) Processing*

Thermomechanical treatment (TMT) processing combines plastic deformation, heating and cooling in different sequences in order to increase strength and toughness simultaneously, by the formation of high density of structural defects. This process is an established strategic method for improving the mechanical properties of advanced high strength steels (AHSS) and further metallic alloys through the control of microstructure [28]. This method was invented in 1950 for C-Mn steel plates and has become an important technique in the process design for steel products, such as plates, sheets, strips, beams bars, wires, pipes and rails [28–33]. For AHSS, the combination of microalloying and suitable TMT may provide an efficient approach to control austenite recrystallization. As a result, fine ferrite grains can be

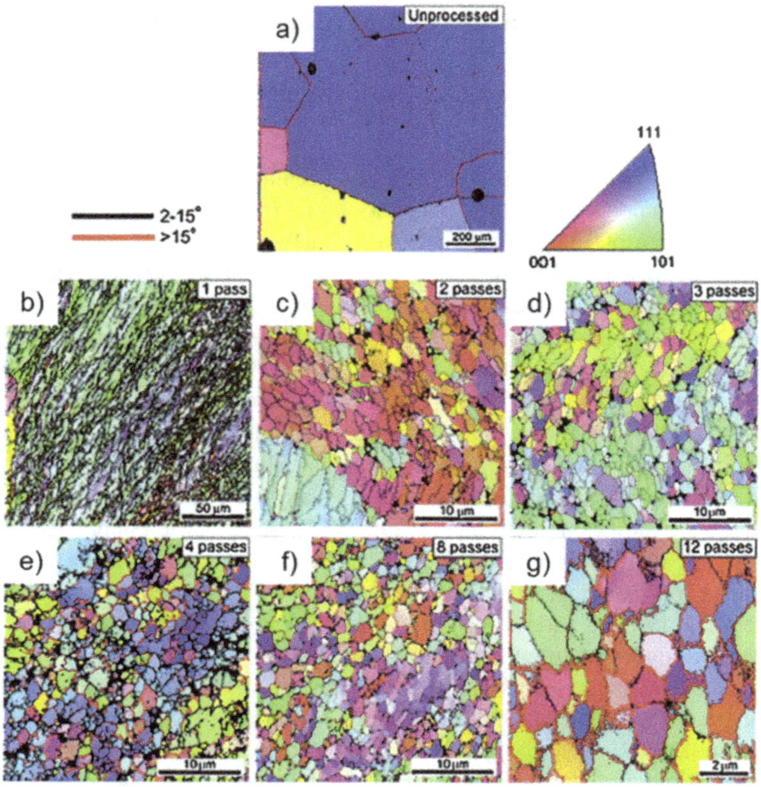

Fig. 2.8 Microstructures in high purity Al in **a** The unprocessed condition; and **b–g** After processing by ECAP through increasing number of passes [27]

obtained by the refined mother phase of austenite during controlled rolling in combination with the flowing accelerated cooling. The aim of TMT for the first generation of AHSSs is to moderately refine the microstructure, in combination with the control of microstructure constituents, phase morphology and precipitation. The TMT is still challenging for the second generation AHSS, due to the elevated alloying additions that are required to stabilise austenite at room temperature. The third generation of AHSS is currently focused on heat treatment, because the microstructural characteristics needed for the steel are primarily achieved by controlling post-deformation heat treatment conditions [34]. So far, only limited articles addressed the rolling/deformation process in the TMT of the third generation [28]. To tailor microstructure and mechanical properties of complex intermetallic high entropy alloys using TMT was investigated by Sunkari et al. [35]. It was figured out that upon annealing the deformed nano-lamellar microstructure transformed into the recrystallized equiaxed ultrafine microstructure. The alloys show retention of the deformation texture components upon annealing, which indicates the absence of preferential nucleation and growth.

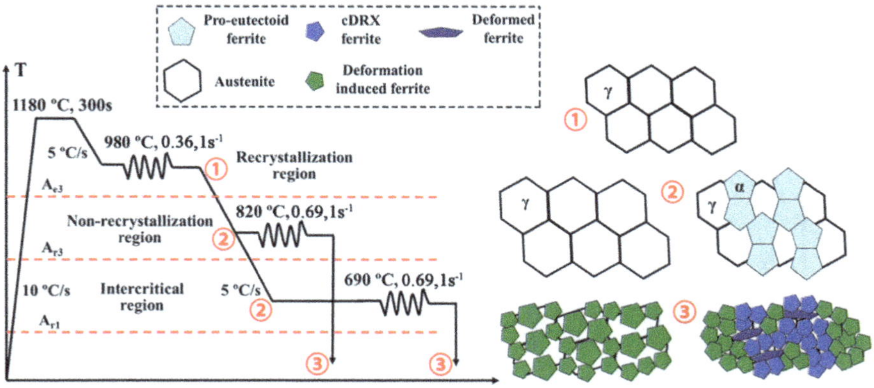

Fig. 2.9 Schematic illustration of grain refinement via advanced thermomechanical processing in commercial low-carbon microalloyed steel (0.04 C, 0.28 Si, 1.54 Mn, 0.011 P, 0.002 S, 0.062 Nb, 0.014 Ti, 0.14 Mo, 0.29 Ni, Cr: 0.25, Fe: bal. (wt.%)) [36]

The advanced thermomechanical processing (ATP) technology is an effective way to acquire the excellent comprehensive property with low cost by achieving the microstructure refinement down to nano-scale [36]. Grain refinement mechanism of the ATP is realized by combining the following phenomena: (i) dynamic recrystallization (DRX) of austenite; (ii) extended phase transformation of the deformed-austenite with the abundant deformation bands; (iii) deformation-induced ferrite transformation (DIFT) by deformation in the low-temperature interval of the austenite non-recrystallization region; and (iv) DRX or dynamic recovery (DRV) of ferrite [36]. Figure 2.9 exhibits the ATP conditions of a commercial low-carbon microalloyed steel. The effective grain size, the grain enclosed by high angle grain boundaries (HAGBs), of the samples before deformation at 820 and 690 °C are 6.4 and 4.3 μm, respectively [36]. The effective grain size of the samples after 0.69 strains at 820 and 690 °C decreases to 1.55 and 1.10 μm, respectively [36]. These increased α/γ interfaces as the nucleation site could accelerate DIFT by repeated nucleation at both the pre-existed and post-formed interfaces. As a result, the DIFT process is enhanced during intercritical deformation at 690 °C [36].

2.2.3 Solid Reaction

Austenite-reverted transformation (ART) is advantageous in tailoring the microstructure for multi-phase metastable steels in terms of morphology, phase constitution, grain size, the volume percentage of austenite, etc. This phase transformation proceeds in accompany with the redistribution of alloying elements to stabilize metastable phase, which is commonly performed for medium-manganese steel (MMnS) grade. MMnS, belonging to the third generation of AHSS, is a type of

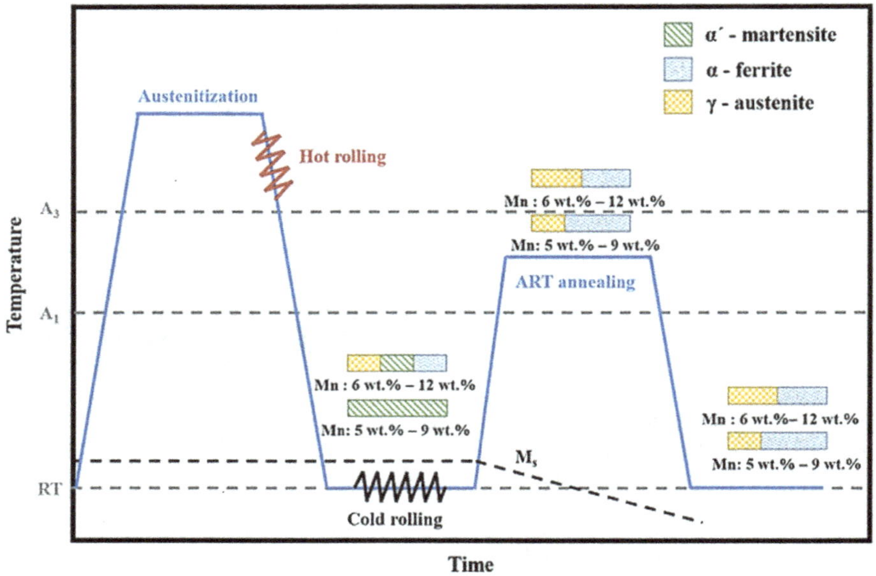

Fig. 2.10 Schematic illustration of ART annealing of MMnS [37]

promising structural materials to be used for various applications, due to their excellent mechanical properties and low alloying costs. In MMnS, considerable amount of austenite can be retained at room temperature by enrichment of carbon and manganese atoms. To achieve the targeted duplex microstructure, the ART process is schematically shown in Fig. 2.10 [37]. The heat treatment route consists of pre-austenitization, hot and cold rolling, and intercritical annealing. The steel strip is heated above A_3 temperature to achieve a fully austenitic microstructure and relatively uniform element distribution. Subsequently, after the hot rolling and cold rolling processes, a fully deformed martensitic microstructure is obtained [37].

During the intercritical annealing, carbon and manganese atoms are enriched at preferential nucleation sites to form nano-sized austenite, *e.g.* martensite lath boundaries, carbides. These manganese and carbon atoms partition from parent martensitic/ferritic phase into austenite grains, stabilizing the reverted austenite. The amount, grain size and morphology of reverted austenite are dependent on the initial microstructure, alloying elements and annealing conditions. At relatively low intercritical temperatures, the reverted austenite possesses good thermal stability, owing to the significant amount of austenite stabilizing elements, and can be retained at room temperature. According to the thermodynamics, the austenite formed at high intercritical temperatures reveals lower amount of carbon and manganese and higher volume fraction. The reverted austenite with insufficient stability can be converted back to martensite during final cooling to room temperature because the stability of the reverted austenite decreases with the increased annealing temperature [37].

Quenching and partitioning (Q&P) is a novel process to produce high-performance steel by controlling carbon partitioning and phase transformation. The Q&P process is schematically shown in Fig. 2.11, which consists of austenitization, interrupted quenching, and partitioning tempering [38]. A fully austenitic microstructure is quenched to an intermediate temperature between the martensite start (M_S) temperature and martensite finish (M_f) temperature. At quenching temperature, martensite transformation is incomplete, resulting in a mixture of quenched martensite and untransformed austenite. Subsequently, an isothermal partitioning step is performed at either identical temperature or the elevated temperature, where carbon atoms further partition from carbon-supersaturated martensite into the retained austenite. As the result, the thermal stability of these austenite is enhanced and can be retained in the final microstructure [38].

A new generation of bainitic steels is designed with the phase transformation at low temperature (200–350 °C), leading to a nano-scale structure, known as NANOBAIN. The microstructure consists of slender ferritic crystals with a size of 20–40 nm. Low transformation temperatures are associated with fine microstructures, which in turn possess high strength and toughness. The theory described above can be used to develop steels that transform to bainite at temperatures as low as 125 °C. Figure 2.12 shows the processing of NANOBAIN by isothermal annealing at low temperature and the respective austenite fraction [39]. The retained austenite fraction is expected to increase for the higher transformation temperature because less bainite forms; this is in contrast to the situation with low-carbon alloys, where a larger fraction of bainite favours the retention of austenite because of the partitioning of carbon into the austenite [39].

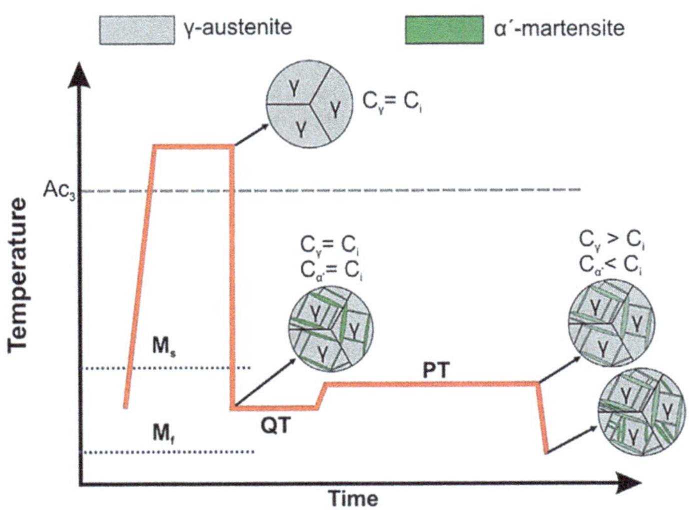

Fig. 2.11 Schematic illustration of the Quenching and Partitioning (Q&P) process. The light and dark green colours illustrate the martensite with different carbon contents [38]

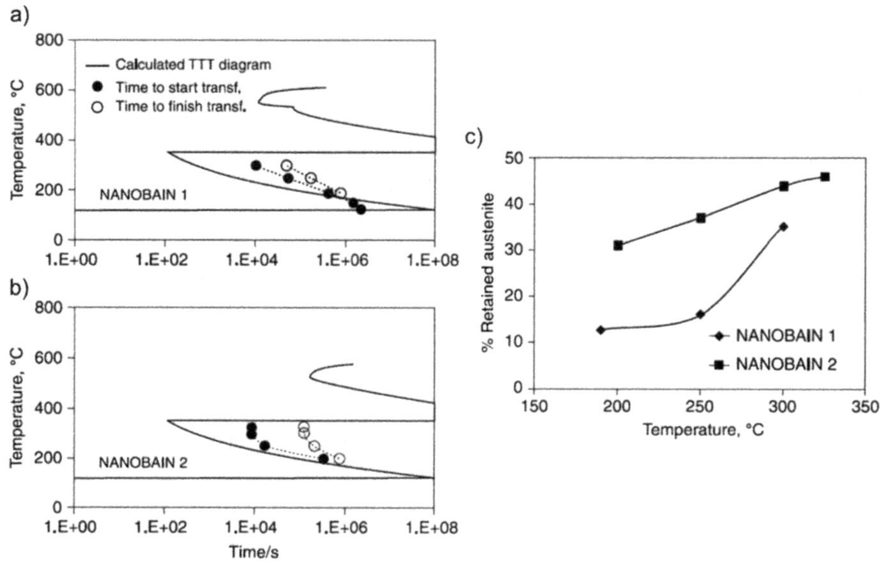

Fig. 2.12 Calculated TTT diagrams for the initiation of transformation, and measured times for the commencement (filled points) and termination of reaction (open circles): **a** NANOBAIN 1 (0.8C-1.6Si-1.9Mn-1.3Cr-0.3Mo-0.1V); **b** NANOBAIN 2 (1.0C-1.5Si-1.9Mn-1.3Cr-0.3Mo-0.1V); **c** X-ray experimental data on volume fraction of retained austenite [39]

2.3 Properties of Nanostructures

For diverse service environments, different combinations of mechanical properties are tailored to achieve the safety and economic efficiency simultaneously. High strength and superior ductility are long pursued in the steelmaking industry because steels are the mostly used structural materials [40–43]. The mechanical properties of materials are highly correlated with the grain size and crystalline size [44]. In polycrystalline materials, adjacent grains usually do not possess the same slip system for dislocations, which are therefore hindered to cross from one grain to another and are subsequently piled-up at the grain boundary [45]. In brittle materials, the propagation of cracks may lead to a component failure. The crack propagation is described by the Griffith model [46]. From this mechanism, it is evident that a material with a maximum crack length of several nm requires very high stress to cause a brittle failure. Although the smaller grains in the materials suggest a higher strength, a reversed Hall–Petch effect where softening in the bulk nano-phase materials with small grain sizes (<10 nm) is observed [47–49]. For a crystallite size of about 5 nm, there are 40–50% of all atoms situated in grain boundaries, where these atoms are not as strongly bounded as atoms inside the crystals. Simulation work shows that in this case the deformation is caused by a large number of small sliding events at the grain boundaries [48–51].

The concept of ultrafine grain refers to reducing the grain size until at least double strength is achieved. The cold drawing and other SPD methods also may lead to grain refinement (<100 nm) in metals because of the formation of high-angle grain boundaries by rearrangement of intensive dislocations. The cold-drawn high-C steel wires of approximately eutectoid composition are reported to exhibit strength up to 5 GPa–6 GPa [52]. Apart from the Hall–Petch effect, various strengthening mechanisms may affect. In newly developed deformed and partitioned (D&P) steel, the microstructure contains metastable austenite grains embedded in a highly dislocated martensitic matrix. The high dislocation density in the D&P steel not only increases the yield strength by dislocation forest hardening but also enables a large ductility by the glide of existing mobile dislocations and by the controlled release of the TRIP effect [53].

An example of extremely high strength obtained in ZnO nano-wires is shown in Fig. 2.13. By employing *in-situ* electron microscope techniques, the real-time atomic-scale deformation of nano-wires under load can be observed, providing a powerful tool for understanding the mechanical behaviours. The yield or fracture strength of a bulk material is significantly lower than the respective theoretical values predicted using the atomic-bonding energy, due to the existence of defects or flaws in the material. The nano-wire materials utilize the strategy to strengthen a material by reducing defects or flaws [54]. The measured mechanical properties of ZnO nano-wires are shown in Fig. 2.13a. The fracture strength and ultimate strength of nano-wires can reach up to 8 GPa-12 GPa and are related with the surface area and diameter.

Barai et al. [55] investigated the creep resistance in nano-crystalline solids by molecular dynamics. A significant fraction of atoms tends to enrich in the grain boundary region. The simulations reveal that there are two distinct regions, one the hard grain interior and the outter softer grain-boundary zone. The grain interior is crystalline whereas the grain boundary (GB) zone is largely amorphous [55]. As a consequence, the material behaves like a composite. A composite model is developed to examine the creep resistance [55]. With this model, the critical grain size is identified, at which the material has maximum creep resistance. An increase of the creep resistance is attributed to the decrease of grain size through Hall–Petch

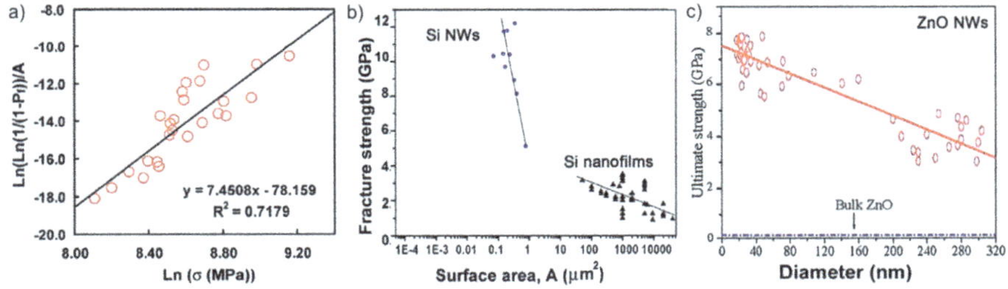

Fig. 2.13 **a** Fracture strength obtained from *in-situ* TEM tensile test plotted against the surface area of ZnO NWs; **b** Fracture strength obtained from *in-situ* SEM tensile test plotted against the side surface area of Si NWs and nano-films; **c** Fracture strength obtained from AFM three-point bending test plotted as against the diameter of ZnO NWs [54]

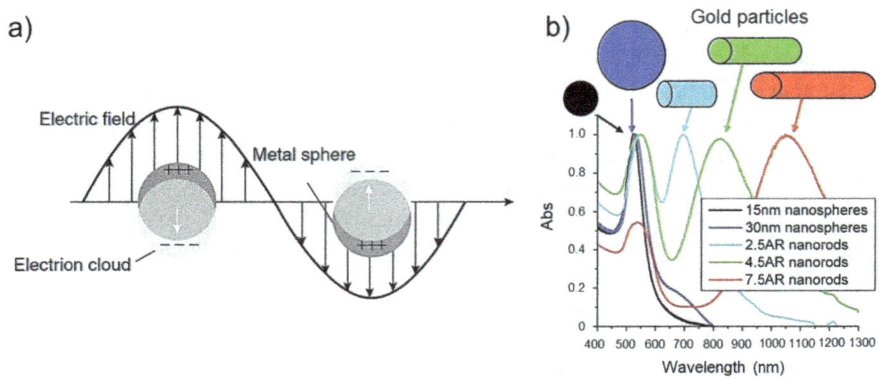

Fig. 2.14 **a** Schematic of the localised surface plasma resonance (LSPR) on the surface of the metallic nanostructure as a result of coherent interaction of electrons in the conduction band with light [1]; **b** Representation of absorbance–wavelength curve for different sizes and shapes of Au nano-particles [57]

effect, but a continuous decrease of grain size would increase the presence of the softer grain boundary and this, in turn could result in the softening effect for the nano-crystalline solid [55].

The materials exhibit specific physical properties in the service environment subjected to a certain stimulus [56]. These properties include electric, magnetic and optical functional properties. Figure 2.14a schematically shows the mechanisms of resultant optical properties. It is well established that the peak wavelengths of the spectrum of different nano-particles are dependent on the size, the shape and the particle interspacing as well as their own dielectric properties (Fig. 2.14b) [57], making it possible to tailor the desired optical properties.

2.4 Applications of Nanostructures

Nano-technology enables sustainable, comfortable, safe and environmentally friendly solutions for automotive sector [58, 59]. The automotive industry benefits from nanomaterials in several more domain like frames and body parts, engines and powertrain, tribological aspects, paints and coatings, suspension and breaking systems, lubrication, tires, exhaust systems [60]. The approach of utilizing the nanofluids leads to a 100% increase in liquid thermal conductivity [60, 61]. Steels have been the materials of choice in the automotive industry since the first vehicles were manufactured. The advantages of steels include the readily availability, low cost, and high recyclability. Recently, the demands for increased safety and fuel economy have promoted the development of AHSS. Owing to the excellent balance of strength and ductility, third generation of AHSS is designed and developed. The ultrafine-grained medium-Mn steels, Q&P steels and carbide-free bainitic steels belong to this grade, showing the great potential in automobile application. Maintaining the ductility above 30%, the strength is increased up to 1.5 GPa. The high strength is attributed to reduced grain size, nano-precipitates and nano-plates. The medium-Mn steel Fe-5Mn-2.5Al-0.2C for automotive application is systematically studied [62]. The cold-rolled materials are intercritically annealed in ferrite–austenite region and air-cooled to 400 °C, then quenched to room temperature. The microstructure is a mixture of ultrafine-grained (UFG) ferrite and austenite (1 μm or less in size) and reveals the tensile strength over 1250 MPa [62].

The composite materials are developed in automotive industry for the aim of weight reduction and fuel reduction. Tesla used a high power to weight ratio tanks and sink of lightweight carbon fibre/epoxy composite. The Evora produced by Lotus has an aluminium chassis along with a composite roof and stressed composite body panels [60]. A study of Fontaras and Samaras [63] showed that roughly 13% of CO_2 emission might be reduced by 10% reduction of the vehicle mass in conjunctions with 10% aerodynamic optimization, or 20% reduced tires rolling resistance in conjunction with 7.5% increase in average powertrain efficiency. Volkswagen used long-fibre reinforced thermoplastic (LRFT) in the dashboard of the Golf Plus, which leads to an elimination of metal and processing steps resulting in a weight reduction of about 8 kg [64].

Automobile tires contain materials such as carbon black and silica fillers, nano-clay, carbon nano-tubes and graphene as well as oils and other additives, surrounded by a matrix of polybutadiene and styrene-butadiene rubber. These nano-scale additives are used as pigments and reinforcing agents during tire manufacturing. Nano-additives are incorporated into the rubber before vulcanization [65]. Dispersion of nano-scale carbon black inclusions increases the tensile strength and wear resistance of tires due to the chemical double bond between the carbon black and the rubber. Nano-size silicaparticles can enhance the rolling resistance and the wet traction [66]. These particles are used to increase the friction between the surface of the tire and the road for improved rolling resistance and wet tire traction. Nano-additives also improve the lifetime of tires [67].

Nano-glass coatings (25–300 nm) like indium-tin oxide or titanium dioxide have dramatically increased the quality of glass surfaces. These coatings make a glass surface hydrophobic and thus preventing water deposits on the glass. Furthermore, the glass is protected from micro-scratches caused by environmental factors [68]. Polymeric coatings are used to protect the surface of materials against environmental influences (UV-light, moisture, oxygen, pollutants). These organic films include polyurethane, resin, polyester, polyamide and epoxy [69]. Under longer exposure to environmental effects, these coating experience physical and chemical deterioration. To prevent environmental effect on organic coatings, a number of different nano-particles like nano-clay, TiO_2, graphene, SiO_2, etc. are dispersed in the primers and top coatings and subsequently sprayed on the surface of automobile [70–72]. Figure 2.15 displays the $NiTiO_3$ (NTO) and nickel-rich layered oxides (NLO) nano-coating process [73]. The NTO nano-coating layer is bond to the NLO materials. Via the HRTEM inspection, the atomic structure at the interface was analysed (Fig. 2.15b), where the atomically sharp crystal structure is revealed (Fig. 2.15c).

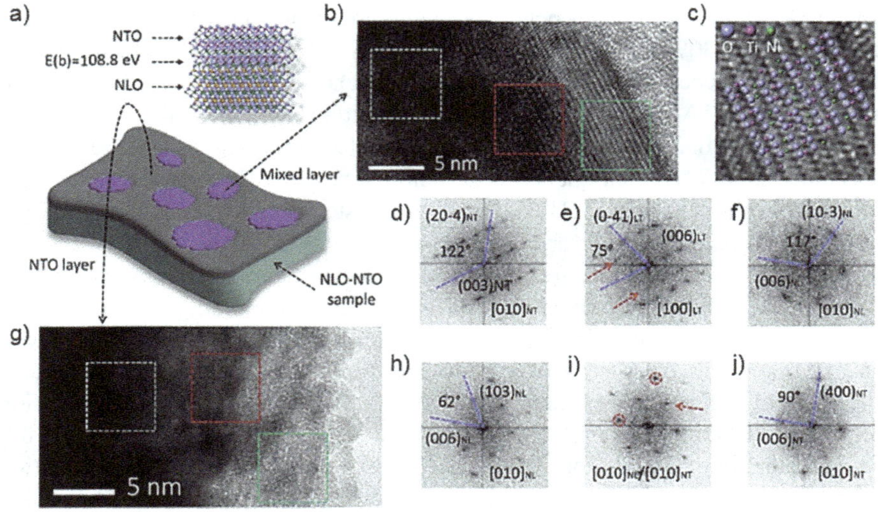

Fig. 2.15 **a** Schematic illustration of the multifunctional NTO nano-coating; **b** The interfacial structures of the mixed coating layer at the NLO-NTO surface; **c** A HRTEM image showing a detailed atomic-structure of the NTO nano-coating; **d**, **e** and **f** are the FFT patterns of the green, red and cyan square marked regions in (**b**), respectively; **g** The interfacial structure of the NTO nano-coating; **h**, **i**, and **j** are the FFT patterns of cyan, red, and green square marked regions in (**g**) [73]

After the Fourier transform, the patterns clearly illustrate the presence of bidirectional diffusion of the TiO_2 and respective crystalline relationship [73]. Depositing TiO_2 nano-particles increase the corrosion protection [74]. Alumina thin films have unique mechanical properties and corrosion resistance. As a result, they have been implemented in many industrial fields such as surface passivation, gas diffusion barriers, antireflection layers, etc. [75]. Ruhi et al. [76] reported a better corrosion resistance of 9Cr-1Mo ferric steel coated with nanostructured sol–gel alumina nano-coating compared to the uncoated steel in a NaCl solution. The nano-coating of alumina was deposited on 100Cr6 carbon steel with the atomic layer deposition method at 160 °C. It is revealed that the thickness of the coating needs to be higher than 10 nm to prevent corrosion. Such thickness is required to seal defects resulting from the initial stages of layer growth [77].

2.5 Nano-Engineering of Metallic Materials

Metallic materials can be considered as one of the most important engineering materials. Nowadays, the application of metallic materials extends from structural materials further towards functional materials and biomaterials. This development trend delivers the high-demanding requirements in properties which can hardly be achieved at microscale by conventional technology. Therefore, the design concept of modern materials is down to the nano-scale. Nano-engineering is proven to be the effective approach guiding us to design the materials at the nano-scale or even at the atomic level for the following several decades in a wide spectrum of industry sectors, including manufacturing, transportation, information technology, energy, electronics and healthcare. The fabrication and testing technologies have boosted in the recent years, facilitating the materials development. Novel devices at the micro- and nano-scale will thus become increasingly important. Technological developments over last few decades have introduced new and powerful techniques into the field of nano-mechanical testing of materials enabling the scale of experiments to be drastically decreased in terms of applied loads, relative displacements and contact pressures. Improved resolution, calibration and efficiency of existing techniques and development of new tools have allowed highly localised and accurate mechanical property information to be obtained. In order to investigate the mechanics behaviour at nano-scale, the great importance of sample preparation should be emphasized.

The focused ion beam (FIB) technique has been known for over three decades, making it the most routine and successful approach to fabricate samples at nano-scale. Owing to the FIB milling process, micro-sized pillars require correction of the measurements of noncircular cross section area for an accurate calculation of the applied stresses during compression (Fig. 2.16a) [78]. The development of FIB and nano-pillar compression test in combination with scanning electron microscopy (SEM) and TEM have been employed to associate hierarchical microstructures with mechanical properties (Fig. 2.16d). Similar as the micro-/nano-pillar, cantilever specimen can be also processed by FIB technique (Fig. 2.16b). The quantitative, accurate modulus values can be extracted from small scale by this technique [79]. Atomic force microscopy (AFM) enables facilitating the surface diagnostics and analysis. The local geometric information, electrical information and mechanical properties of the substrate surface can be obtained. Furthermore, AFM allows to form the nano-scale structures on the surface of solid, which also follow the concept of nano-engineering [80]. The AFM tip is usually fabricated by FIB (Fig. 2.16c). The objective of a nano-indenter is to transfer a controlled load to an accurately shaped indenter that is in contact with the specimen surface. The testing principle is that the load can be applied from an expanding piezoelectric element, movement of a coil in a magnetic field or electrostatically. Mechanical properties, such as hardness and Young's modulus of elasticity, can be determined on micro- to nano-scales using the AFM and a depth-sensing indentation system used in conjunction with an AFM (Fig. 2.16e, f). Additionally, it can provide information on cracking, creep, strain hardening, phase transformation, fracture toughness and energy absorption.

With the comprehensive understanding of the nanostructures as well as the processing technology, the properties can be tailored according to the service environment. Different nano-engineering concepts are developed to improve the materials' performance, *e.g.* precipitation engineering (Chap. 5), interface engineering (Chap. 6) and short-range ordering (SRO) engineering (Chap. 7). These new material design concepts at nanometre scale speed up the development of novel metallic materials with extraordinary properties.

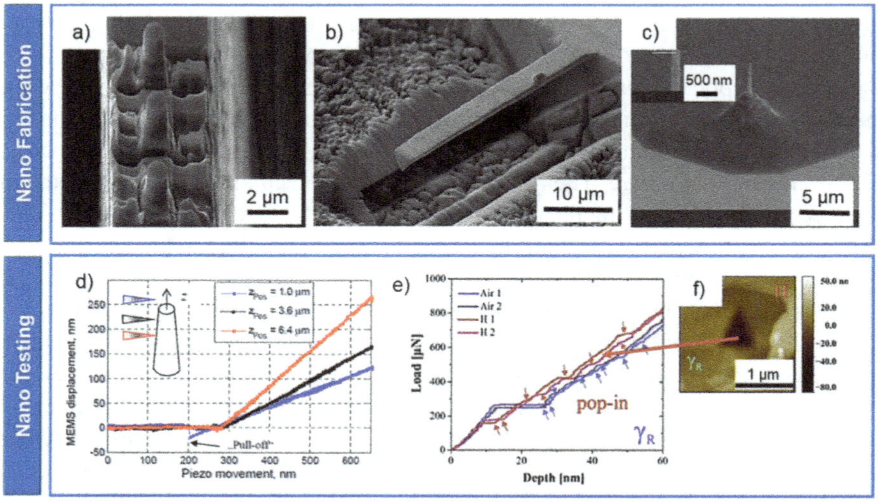

Fig. 2.16 Nano-fabrication and nano-testing techniques: FIB preparation of **a** Nano-pillar [78]; **b** Micro cantilever [79]; and **c** Atomic force microscopy (AFM) tip [80]; **d** Nano-bending test [81]; **e** Dislocation pop-in behaviour characterized by nano-indentation [82] and **f** The corresponding AFM micrograph [82]

References

1. Cheng X (2014) Nanostructures: fabrication and applications. In Feldman M (ed), Nanolithography, Cambridge, Woodhead Publishing
2. Sun B, Ma Y, Vandresse N, Varanasi RS, Song W, Bocher P, Ponge D, Raabe D (2019) Macroscopic to nanoscopic *in-situ* investigation on yielding mechanisms in ultrafinegrained medium Mn steels: role of the austenite-ferrite interface. Acta Mater 178:10–25
3. Li Y, Raabe D, Herbig M, Choi PP, Goto S, Kostka A, Yarita H, Borchers C, Kirchheim R (2014) Segregation stabilizes nanocrystalline bulk steel with near theoretical strength. Phys Rev Lett 113:106104
4. Li Y, Choi P, Borchers C, Westerkamp S, Goto S, Raabe D, Kirchheim R (2011) Atomic-scale mechanisms of deformation-induced cementite decomposition in pearlite. Acta Mater 59:3965–3977
5. Chen S, Cao Z, Zhang M, Xu H, Dzugan J, Cao W (2018) The hot rolled duplex laminated steel presenting both high strength and high toughness, IOP Conf Ser Mater Sci Eng A 461:012011
6. Wang M, Tasan C, Ponge D (2015) Nanolaminate transformation-induced plasticity-twinning-induced plasticity steel with dynamic strain partitioning and enhanced damage resistance. Acta Mater 85:216–228
7. Guo W, Jägle EA, Choi PP, Yao J, Kostka A, Schneider JM, Raabe D (2014) Shear-induced mixing governs codeformation of crystalline-amorphous nanolaminates. Phys Rev Lett 113:035501
8. Caballero F, Rementeria R (2019) Understanding mechanical properties of nano-grained bainitic steels from multi-scale structural analysis. Metals 9:426
9. Jiang S, Hui W, Wu Y, Liu X, Chen H, Yao M, Gault B, Ponge D, Raabe D, Hirata A, Chen M, Wang Y, Lu Z (2017) Ultrastrong steel via minimal lattice misfit and high-density nanoprecipitation. Nature 544:460–464
10. Podgornik B, Torkar M (2015) Improving properties of spring steel through nano-particles alloying. Mater Sci Eng A 638:183–189

11. Sun L, He X, Lu J (2018) Nanotwinned and hierarchical nanotwinned metals: a review of experimental, computational and theoretical efforts, *npj Comput*. Mater 4:6
12. Mayer J, Beigmohamadi M, Lipinska-Chwalek M, Wittig JE (2016) Atomic level characterization of novel hardening mechanisms in high-Mn-steels. Microsc Microanal 22:1928–1929
13. Roesner H, Boucharat N, Padmanabhan K, Markmann J, Wilde G (2010) Strain mapping in a deformation-twinned nanocrystalline Pd grain. Acta Mater 58:2610–2620
14. Bridgman P (1952) Studies in large plastic deformation and fracture, 1st edn. Harvard University Press, New York
15. Valiev R, Estrin Y, Horito Z, Langdon T, Zehetbauer N, Zhu Y (2006) Producing bulk ultrafine grained materials by severe plastic deformation. J Met 58:33–39
16. Segal V (2018) Review: modes and prcocesses of severe plastic deformation (SPD). Materials 11:1175
17. Nutting J (1971) Some aspects of structure-property relationship in materials. In Electron Microscopy and Structure of Materials, Berkeley, University of California Press
18. Erbel S (1979) Mechanical properties and structure of extremely hardened copper. Met Technol 6:482–486
19. Korbel A, Richert M, Richert J (1981) The effect of very high accumulative deformation on structure and mechanical properties of aluminium. In Proceedings of the 2nd Riso Symposium Deformation of Polycrystalls: Mechanisms and Structures, Riso, Denmark
20. Sanders I, Nutting J (1984) Deformation of metals to high strains using combination of torsion and compression. Met Sci 18:571–575
21. Rigney D, Naylor M, Davikar S, Ives L (1986) Low energy dislocation structures caused by sliding and by partical impact. Mater Sci Eng A 81:409–425
22. Smirnova N, Levit V, Pilugin V, Kuznetsov R, Davydova L, Sazonova V (1986) Microstructure evolution of f.c.c. monocrystalls under large plastic deformation. Phys Met Metallogr 61:1170–1177
23. Segal V, Reznikov V, Drobyshevski A, Kopylov V (1981) Plastic working of metals by simple shear. Russ Metall 1:99–105
24. Segal V, Reznikov V, Kopylov V, Pavlik D, Malyshev V (1994) Processes of plastic structure formation in metals. Sci Eng BSSR 231
25. Estrin Y, Murashkin M, Valiev R (2010) Ultrafine grained aluminium alloys: processes, structural features and properties. In Lumley R (ed) Fundamentals of aluminium metallurgy: production, processing and applications, Cambridge, Woodhead
26. Estrin Y (2013) Extreme grain refinement by sever plastic deformation: a wealth of challenging science. Acta Mater 61:782–817
27. Huang Y, Langdon T (2013) Advances in ultrafine-grained materials. Mater Today 16:85–93
28. Zhao J, Jiang Z (2018) Thermomechanical processing of advanced high strength steels. Prog Mater Sci 94:174–242
29. Gong P, Palmiere EJ, Rainforth WM (2015) Dissolution and precipitation behaviour in steels microalloyed with niobium during thermomechanical processing. Acta Mater 97:392–403
30. Cao J, Yan J, Zhang J, Yu T (2015) Effects of thermomechanical processing on microstructure and properties of bainitic work hardening steel. Mater Sci Eng A 639:192–197
31. Nezakat M, Akhiani H, Hoseini M, Szpunar J (2014) Effect of thermo-mechanical processing on texture evolution in austenitic stainless steel 316L. Mater Charact 98:10–17
32. Kim Y, Song S, Seo S, Hong S, Lee C (2013) Development of Ti and Mo micro-alloyed hot-rolled high strength sheet steel by controlling thermomechanical controlled processing schedule. Mater Sci Eng A 565:430–438
33. Kong X, Lan L, Hu Z, Li B, Sui T (2015) Optimization of mechanical properties of high strength bainitic steel using thermo-mechanical control and accelerated cooling process. J Mater Process Technol 217:202–210
34. Poliak E, Bhattacharya D (2014) Aspects of thermomechanical processing of 3rd generation advanced high strength steels. Mater Sci Forum 783–786:3–8

35. Sunkari U, Reddy S, Rathod B, Kumar D, Saha R, Chatterjee S, Bhattacharjee P (2020) Tuning nanostructure using thermo-mechanical processing for enhancing mechanical properties of complex intermetallic containing CoCrFeNi2. 1Nbx high entropy alloys. Mater Sci Eng A 769:138489
36. Li B, Liu Q, Jia S, Ren Y, Wang B (2018) Fabricating ultrafine grain by advanced thermomechanical processing on low-carbon microalloyed steel. Scr Mater 152:132–136
37. Ma Y (2017) Medium-manganese steels processed by austenite-reverted-transformation annealing for automotive applications. Mater Sci Technol 33:1713–1727
38. Bleck W, Guo X, Ma Y (2017) The TRIP effect and its application in cold formable sheet steels. Steel Res Int 88:1700218
39. Caballero F, García-Mateo C (2011) The processing of nanocrystalline steels by solid reaction. In Whang SH (ed) Nanostructured metals and alloys, Cambridge, Woodhead Publishing
40. Guo J, Sun Y, Zhang X, Tang L, Liu H (2014) FePt nanoalloys anchored reduced graphene oxide as high-performance electrocatalysts for formic acid and methanol oxidation. J Alloys Compd 604:286–291
41. Callister W (2007) Materials science and engineering: an introduction, 7th edn. John Wiley & Sons, New York
42. Sedlmayr A, Erik B, Gianola D, Richter G, Mönig R, Kraft O (2012) Existence of two twinning-mediated plastic deformation modes in Au nanowhiskers. Acta Mater 60:3985–3993
43. Mordehai D, Martin G (2011) Enhanced annealing of the dislocation network under irradiation. Phys Rev B 84:014115
44. Siegel R (1994) Nanophase materials. In Trigg G (ed) Encyclopedia of applied physics, Weinheim, VCH Publishers
45. Gottstein G (2004) Physical foundation of materials science. Springer, Berlin
46. Anderson J, Leaver K, Rawlings R, Alexander J (1990) Material science, 4th edn. Chapman & Hall, London
47. Yip S (1998) The strongest size. Nature 391:532–533
48. Schiotz J, Di Tolla F, Jacobsen K (1998) Softening of nanocrystalline metals at very small grain sizes. Nature 391:561–563
49. van Swygenhofen H, Caro A (1997) Plastic behavior of nanophase Ni: a molecular dynamics computer simulation. Appl Phys Lett 71:1652
50. van Swygenhofen H, Caro A, Farkas D (2001) Grain boundary structure and its influence on plastic deformation of polycrystalline fcc metals at the nanoscale: a molecular dynamics study. Scr Mater 44:1513–1516
51. Arzt E (1998) Size effects in materials due to microstructural and dimensional constraints: a comparative review. Acta Mater 46:5611–5626
52. Zhang X, Godfrey A, Huang X, Hansen N, Liu Q (2011) Microstructure and strengthening mechanisms in cold-drawn pearlitic steel wire. Acta Mater 59:3422–3430
53. He B, Hu B, Yen H, Cheng G, Wang Z, Luo H, Huang M (2017) High dislocation density-induced large ductility in deformed and partitioned steels. Science 357:1029–1032
54. Wang S, Shan Z, Huang H (2017) The mechanical properties of nanowires. Adv Sci 4:1600332
55. Barai P, Weng G (2008) Mechanics of creep resistance in nanocrystalline solid, Acta Mech
56. Kassing R, Petkov P, Kulisch W, Popov C (2006) Functional properties of nanostructured materials, 1st edn. Springer, New York
57. Eustis S, El-Sayed M (2006) Why gold nanoparticles are more precious than pretty gold: noble metal surface plasmon resonance and its enhancement of the radiative and nonradiative properties of nanocrystals of different shapes. Chem Soc Rev 35:209–217
58. Presting H, Koning U, Mater J (2003) Future nanotechnology developments for automotive applications. Mater Sci Eng C 23:737
59. Malani A, Chaudhari A, Sambhe R (2016) A review on applications of nanotechnology in automotive industry. IJMME 10:36–40
60. Coelho M, Torrao G, Emami N, Gracio J (2012) Nanotechnology in automotive industry: research strategy and trends for the future—small objects, big impacts. J Nanosci Nanotechnol 12:6621–6630

61. Keblinski P, Eastman J, Cahill D (2005) Nanofluids for thermal transport. Mater Today 8:36–44
62. Perlade A, Antoni A, Besson R, Caillard D, Callahan M, Emo J, Gourgues A, Maugis P, Mestrallet A, Thuinet L, Tonizzo Q, Schmitt J (2019) Development of 3rd generation Medium Mn duplex steels for automotive applications. Mater Sci Technol 35:204–219
63. Fontaras G, Samaras Z (2010) On the way to 130 g CO_2/km-Estimating the future characteristics of the average European passenger car. Energy Policy 38:1826
64. Stewart R (2009) Lightweighting the automotive market. Reinf Plast 53:14–21
65. Loyd S, Lave L (2003) Life cycle economic and environmental implications of using nanocomposites in automobiles. Environ Sci Technol 37:3458–3466
66. Lin Y, Lee W, Chao H, Wang S, Tsou T, Chang-Chien G, Tsai P (2008) Approach for energy saving and pollution reducing by fueling diesel engines with emulsified biosolution/biodiesel/diesel blends. Environ Sci Technol 42:3849–3855
67. Hong Q, Gu L, Jie X, Wu H, Lei P (2011) Nanotechnology in automotive industry: research strategy and trends for the future. Adv Mater Res 148:498
68. Nguyen P (2011) The nanotechnology and nanosafety in automotive industry, Master thesis, Wichita State University
69. Asmatulu R, Mahmud G, Hille C, Misak E (2011) Effects of UV degradation on surface hydrophobicity, crack and thickness of MWCNT-based nanocomposite coatings. Prog Org Coat 72:553–561
70. Asmatulu R, Claus R, Mecham J, Corcoran S (2007) Nanotechnology: associated coatings for aircrafts. Mater Sci 43:415–422
71. Zaarei D, Sarabi A, Sharif F, Kassiriha S (2008) Structure, properties and corrosion resistivity of polymeric nanocomposite coatings based on layered silicates. J Coat Technol Res 5:1935–3804
72. Asmatulu R, Khan S, Misak H, Nuraji N (2011) Graphene-based nanocomposite coating on fiber reinforcedcomposites subjected to UV degradation. In Proceedings of the SAMPE Fall Technical Conference, Fort Worth, TX
73. Xu M, Fei L, Zhu S, Lu W, Lai Y, Zhang Z, Lam C, Huang H (2018) Multifunctional $NiTiO_3$ nanocoating fabrication based on the dual-Kirkendall effect enabling a stable cathode/electrolyte interface for nickel-rich layered oxides. J. Mater. Chem. A 6:2643–2652
74. Shen G, Chen Y, Lin L, Lin C, Scantlebury D (2005) Study on a hydrophobic nano-TiO_2 coating and its properties for corrosion protection of metals. Electrochim Acta 50:5083–5089
75. Calle E, Ortega P, von Gastrow G, Martín I, Savin H, Alcubilla R (2016) Long-term stability of Al_2O_3 passivated black silicon. Energy Proc. 92:341–346
76. Ruhi G, Modi O, Singh I (2009) Corrosion behaviour of nano structured sol-gel alumina coated 9Cr-1Mo ferritic steel in chloride bearing environments. Surf Coat Tech 204:359–365
77. Díaz B, Härkönen E, Maurice V, Światowska J, Seyeux A, Ritala M, Marcus P (2011) Failure mechanism of thin Al_2O_3 coatings grown by atomic layer deposition for corrosion protection of carbon steel. Electrochim Acta 56:9609–9618
78. Kuzmin O, Pei Y, De Hosson J (2014) Nanopillar fabrication with focused ion beam cutting. Microsc Microanal 20:1581–1584
79. Gibson J, David A, Roberts S (2014) The micro-mechanical properties of ion irradiated tungsten. Phys Scr 159:014056
80. Ageev O, Kolomiytsev A, Bykov A, Smirnov V, Kots I (2015) Fabrication of advanced probes for atomic force microscopy using focused ion beam. Microelectron Reliab 55:2131–2134
81. Li Z, Gao S, Brand U, Hiller K, Hahn S, Hamdana G, Peiner E, Wolff H, Bergmann D (2019) Nanomechanical characterization of vertical nanopillars using an mems-spm nano-bending testing platform. Sensors 19:4529
82. Lu X, Ma Y, Zamanzade M, Deng Y, Wang D, Bleck W, Song W, Barnousch A (2019) Insight into hydrogen effect on a duplex medium-Mn steel revealed by *in-situ* nanoindentation test. Int J Hydrogen Energy 44:20545–20551

Chapter 3
High Strength Steels

This chapter provides the general introduction to the state-of-the-art development of high strength steels. The fundamentals of the deformation mechanisms and its controlling parameters in high strength steels are addressed. The microstructure-mechanical property relationship in high strength steels, in particular, the second generation and the third generation advanced high strength steels (AHSS) are focused. In the end, own work-related example[1] is given to introduce the prediction of strain hardening behaviours based on deformation mechanism maps. The controlling of strain hardening behaviours are further discussed. Based on the alloy design by using mechanism maps calculation, steels with high strength and high strain hardening transformation-induced plasticity (TRIP)/twinning-induced plasticity (TWIP) are achieved. The mechanical properties are compared with the Cr stainless steels.

3.1 Development of High Strength Steels

High strength steels are widely applied in all sectors in the industry and our daily life. In particular, the developments of advanced automobile industry greatly increase the requirement for high strength materials. For weight saving, emission reduction and improvement of passenger safety, higher strength-weight ratio steels are strongly needed [1]. Three generations of advanced high strength steels are developed to fulfil the requirements. Superior in strength ensures the reduction of sheet thickness, good performance in deformation grant possibility for complex component production. Furthermore, the combination of high strength and ductility indicates the outstanding high-energy-absorption-properties and great attraction in crash-safety application.

Own work-related example links to the publication:

[1] Song et al. [77].

W. Song, *Nano-Engineering of High Strength Steels*,
Topics in Mining, Metallurgy and Materials Engineering,
https://doi.org/10.1007/978-3-031-42967-5_3

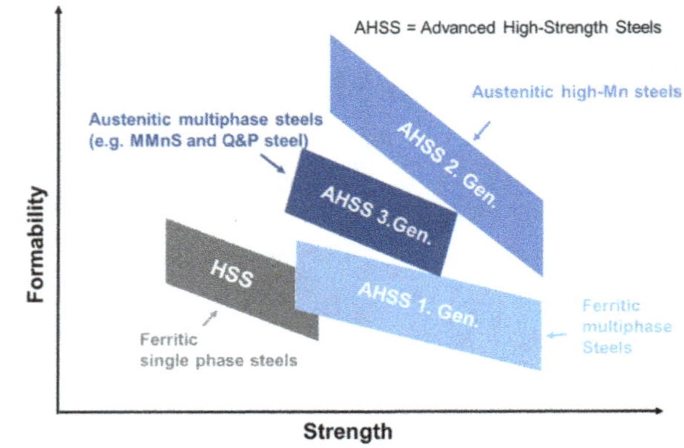

Fig. 3.1 The strength and formability of the different generations of high strength steels [1]

Figure 3.1 shows the balance of strength and formability of the different generations of high strength steels.

3.1.1 1st Generation Advanced High Strength Steels

The 1st generation AHSS have been developed by creating and modifying additional strengthening phases in ferritic microstructure, like dual-phase (DP) steels, complex-phase (CP) steels, conventional TRIP steels and martensitic steels. The Eco-Index (product of ultimate tensile strength and uniform elongation) of the 1st generation AHSS is about 15 ± 10 GPa%. Utilization of the 1st generation AHSS such as DP steels and TRIP steels in automotive body parts allow up to 25% weight saving and up to \$1.6 kg^{-1} reduction in the cost [1].

DP steels consist of soft ferrite matrix and hard martensite islands. The soft ferrite generally distributes continuously for many DP steel grades up to DP 780, but the ferrite may become discontinuous as the volume fraction of martensite exceeds 50%. Ferrite provides DP steels with excellent ductility, while martensite islands contribute to strength, and the higher the volume fraction of the martensitic phase, the higher the strength of the steels will be. Such microstructural characteristic enables DP steels achieving ultimate tensile strength (UTS) in the range of 500–1200 MPa [1].

The microstructure of CP steels is composed of ferrite/bainite matrix with a small amount of retained austenite, martensite and pearlite. Generally, CP steels possesses very similar alloying elements to that of DP and TRIP steels. Additionally, CP steels contain small quantities of niobium, titanium and/or vanadium to form fine precipitates and realise precipitation strengthening. A wide range of mechanical properties can be obtained with the same chemical compositions by adjusting the volume fractions of the phases in CP steels through control of cooling rate [1].

The microstructure of TRIP steels consists of retained austenite embedded in a ferrite matrix. The amount of retained austenite is in a range of 5–20 vol.%. Hard phases such as martensite and bainite are also present in varying amounts. Upon deformation, the metastable retained austenite progressively transforms to martensite, thus retarding the localized deformation in the stress concentrated regions. Such unique transformation allows for high strain-hardening rate to endure at very high strain levels, and contributes to enhanced ductility, which is known as the "TRIP" effect [2].

3.1.2 2nd Generation Advanced High Strength Steels

The 2nd generation AHSS have been intensively studied since the 2000s. The 2nd generation AHSS are mainly high-manganese steels with austenitic single-phase microstructures. Taking advantage of various deformation mechanisms in high-Mn steels, like dislocation slip (SLIP), TRIP, TWIP, and microband-induced plasticity (MBIP), the 2nd generation AHSS manifest extreme ductility (total elongation up to 60%) with high strength (ultimate tensile strength up to 1100 MPa). The Eco-Index of the 2nd generation AHSS is as high as 60 ± 10 GPa%.

High-Mn TWIP steels are austenitic steels, which can deform by both dislocation gliding and mechanical twinning on the (111) <111> system. TWIP steels contain Mn in the level of 15–30%, which is important to stabilize austenite phase in the room temperature. High-Mn TWIP steels obtain outstanding mechanical properties combining high strength (800–1200 MPa) and ductility (total elongation 60–80%), which reveals ultra-high work-hardening capacity (ECO Index 50–60 GPa%) compared to conventional AHSS (ECO Index <25 GPa%).

MBIP steel was first proposed by Yoo in a high Mn-Al-C lightweight steel (Fe-28Mn-9Al-0.8C in wt.%). MBIP steels are composed of single-phase austenite with orderly arranged nano-scale kappa phase in the matrix. Kappa phase, also called kappa phase, has the chemical composition of (Fe, $Mn)_3AlC_x$, where x is usually smaller than 1. In Fe-Mn-Al-C alloy system, the age-hardenability, as well as superior combination of strength and ductility, is attributed to the precipitation of kappa phase in the matrix.

3.1.3 3rd Generation Advanced High Strength Steels

Although the high strength can be achieved in the 1st generation AHSS, the ductility is limited. The high amount of alloying content in the 2nd generation AHSS raises production cost and creates various drawbacks in the production processes, for instance, Al_2O_3 formation during casting process, severe macroscale segregation and microscale segregation due to high alloying concentration, and challenges to hot dip galvanization because of Al_2O_3 and MnO surface layer formation during continuous annealing etc. [2, 3]. The 3rd generation AHSS has been developed in last two decades to fill the gap between the 1st generation AHSS and 2nd generation AHSS. The quenching and partitioning (Q&P) steel and medium-Mn steel grades are the promising candidates in the 3rd generation AHSS, which possess excellent combination of mechanical properties and alloying cost [2, 3].

The Q&P steels take advantage of low alloying content. Mn, C, Al and Si are the main alloying elements in Q&P steels. Generally, C content is in a range of 0.15–0.3 wt.%. Mn content is controlled in between of 1.5 and 3.0 wt.%. Si and/or Al are alloyed in Q&P steels to inhibit the cementite formation. The microstructure of Q&P steels consists of martensite, retained austenite and ferrite (under partial austenitization condition) [2, 3]. The quenching and partitioning process is expected to generate a fine acicular microstructure of lath martensite interwoven with carbon enriched retained austenite [2]. Compared with the ductility of the conventional martensitic steels, the Q&P steels can offer much higher ductility at the same strength level. The TRIP effect in retained austenite during plastic deformation contributes to the increase in strain hardening, improvement of UTS and uniform elongation [3].

The medium-manganese TRIP steels are processed by intercritical annealing in between A_1 and A_3 temperatures, which is also named austenite-reverted-transformation (ART) annealing [4]. Medium-manganese steels usually contain 4–12 wt.% manganese. The C content is generally around 0.05–0.4 wt.%. Al and Si are around 1.5–3.0 wt.% and 1.2– 2.0 wt.%, respectively. A large amount of retained austenite up to 60 vol.% in the ultrafine-grained (UFG) microstructure results in extraordinary combination of high strength and good ductility in medium-Mn steel, as shown in Fig. 3.2 [5].

3.2 Deformation Mechanisms and Strain Hardening

3.2.1 Transformation-Induced Plasticity (TRIP)

The TRIP effect in steels refers to the deformation-induced transformation of metastable face-centred cubic (fcc) austenite γ into body-centred cubic (bcc) structured α'-martensite and/or hexagonal-close packed (hcp) structured ε -martensite [6–8]. It may contribute simultaneously to the enhancement of strength and ductility, which was reported in meta-stable austenitic stainless steels [9, 10], low carbon TRIP steels [11], high Mn steels [12], medium Mn steels [13] and high entropy steels [14]. The retained austenite in TRIP steels is thermodynamic stable, yet mechanically unstable, and the martensitic transformation is considered to be driven by both stress and strain [15–18]. The martensite transformation is prone to nucleation at the intersections of shear band, ε-martensite, grain boundary, twinning plates, and even dislocation pile-ups [19]. The deformation-induced formation of the hcp-structured martensite as a nucleation and growth process is defined according to the suggested model [7]. The coplanar different Burgers vectors of two $a/2\langle 110\rangle$ dislocations may interact with each other on (111) plane and can from a six-layer hcp region as a nucleus regarding the following equation [7]:

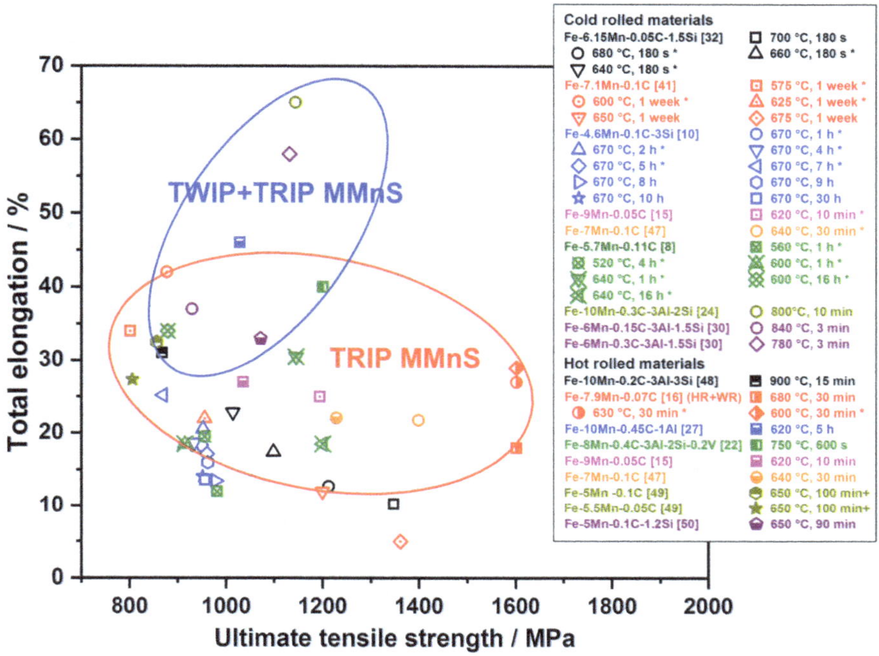

Fig. 3.2 Total elongation versus ultimate tensile strength in medium-Mn steels under different ART-annealing conditions (The stress-strain curves showing discontinuous yielding are marked with *; the stress-strain curves not given in the references are marked with +) [5]

$$\frac{a}{2} <1\overline{1}0> + \frac{a}{2} <10\overline{1}> \rightarrow 3 \times \frac{a}{6} <2\overline{1}\overline{1}> \tag{3.1}$$

Figure 3.3a shows the six-layer hcp nucleus, in which every second plane (D1, D2 and D3) contains an extrinsic stacking fault [20]. The straight lines represent the immobile Shockley dislocations with a length L_{tr} and the bowed out front lines with a radius r represent the mobile Shockley partial dislocations [20]. The martensite transformation induced by deformation also contributes to the enhanced strain-hardening rates [21–23]. The exceptional strain hardening in high-strength Mn steels is attributed to the formation of dislocations and stacking faults during the deformation [23]. The defect structures are found to act as the nucleation sites for the martensite embryos during the strain-induced martensitic transformation. This leads to the high dislocation density in the transformation products and an increased stacking fault population in the austenite matrix. Figure 3.3b shows a TEM micrograph of Fe-12Mn-0.6C (wt.%) steel containing ε-martensite plates within the γ matrix after recrystallization annealing carried out at 800 °C for 100 s [24].

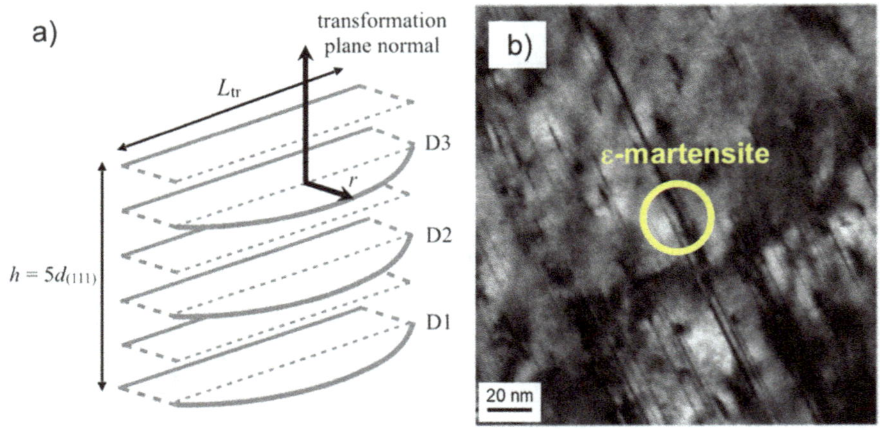

Fig. 3.3 **a** Schematical illustration of a six-layer ε-martensite nuclei, in which the every second plane (D1, D2 and D3) includes an extrinsic stacking fault [20]; **b** TEM micrograph of Fe-12Mn-0.6C steel containing ε-martensite plates within the austenite matrix [24]

3.2.2 Twinning-Induced Plasticity (TWIP)

TWIP effect renders high strength and ductility in austenitic steels owing to the high strain hardening capacity. The high content of Mn and small addition of C, Si, Al in austenitic steels maintain the low stacking fault energy (SFE) between 20 and 40 mJ/m^2, activating the formation of shear-induced mechanical twinning. The calculation of SFE is based on the thermodynamic model, originally proposed by Olson and Cohen [25], which is dependent on the temperature, alloying composition. With the reduction in SFE, the plastic deformation mechanism shifts from dislocation gliding to dislocation gliding plus mechanical twining by toughening the recombination of partial dislocations. As the result, the nucleation of mechanical twinning occurs preferentially at stacking faults. The formed mechanical twin boundaries act as the obstacles for dislocation gliding and exert back stress to successive dislocation motion, contributing to the enhanced strain hardening rate. The TWIP mechanism occurs in stable austenite where the Gibbs free energy $\Delta G^{\gamma\to\varepsilon}$ of the martensitic reaction ($\gamma_{fcc} \to \varepsilon_{hcp}^{Ms}$) is positive of about $\Delta G^{\gamma\to\varepsilon} \approx$ 110–250 J/mol and the stacking fault energy Γ_{fcc} is relatively low, 20–40 mJ/m^2 [26].

The formation mechanisms of the mechanical twinning have been intensively investigated [27–32]. Mahajan et al. [29] introduced a nucleation and growth process for the twin formation. Idrissi et al. [32] compared the different twin formation mechanisms in Fe-Mn-C high-Mn steel and suggested that the model of Cohen et al. [27] and Karaman et al. [28] as most likely to occur. Figure 3.4a schematically illustrates the three-layer twinning nuclei. The straight back lines and bowed out front lines (D1, D2 and D3) respectively stand for the immobile Shockley dislocations with a length L_{tr}, and the mobile Shockley partial dislocations with a radius r [30]. Generation of mechanical twins takes place with the growing of faulted layers into one another [29, 30]. In the model of Cohen et al. [27], it was proposed that pile-up took place when the perfect dislocation faces with the Cottrell-Lomer lock on (111) plane. Dislocation pile-up results in a dissociation of an $a/2[011]$ perfect dislocation into the $a/6[\overline{2}11]$ mobile Shockley partial dislocation on the conjugate slip plane and the $a/3[111]$ Frank sessile partial dislocation due to the stresses at the tip of the pile-up. As seen in Fig. 3.4b, many perfect dislocations dissociate on the (111) plane and the movement of the mobile Shockley partial dislocations (indicated with an arrow) can create several intrinsic stacking faults. Twins are produced by the existence of the intrinsic stacking faults on the adjacent atomic planes [27, 30, 32]. Karaman et al. [28] used the model of Miura, Takamura and Narita (MNT model) to explain the occurrence of twinning in the Hadfield's steel. In the mentioned model, an obstacle like behaviour of Lomer-Cottrell junction against the dislocation motion creates the dislocation pile-up on the primary slip plane and gives rise to a formation of double layer intrinsic stacking fault as a twin nucleus during the plastic deformation. For the evolution of the twinning, the sufficient force should be exerted on the Shockley partials to make them mobile and to separate them from the sessile Frank partial dislocations. Figure 3.4c schematically shows the formation of a double layer stacking fault (twin nuclei) with a critical size of r_c, and the dislocation pile-up with a width of L. The separation of the mobile Shockley partials is enhanced with the increasing L and the decreasing r_c [28].

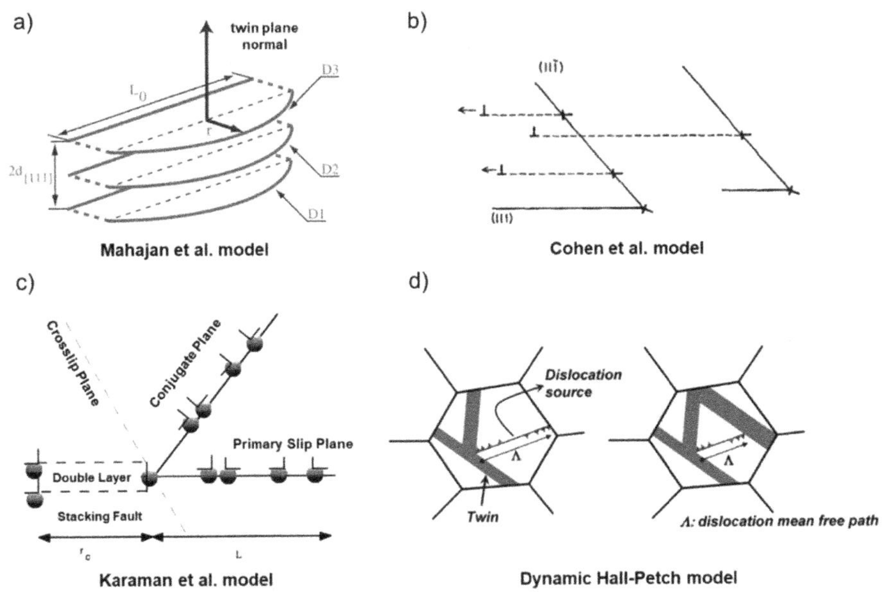

Fig. 3.4 Schematic illustration of **a** A three-layer twin nuclei [20]; **b** A twin formation modal proceeded by the Shockley partial dislocation motion [27]; **c** A mechanical twinning formation model by Karaman et al. [28]; **d** Dynamic Hall-Petch effect [34]

As the stress increases, mechanical twins continuously activate and further subdivide grains into smaller units which is termed as dynamic Hall-Petch effect reported by many authors [31, 33, 34]. As schematically shown in Fig. 3.4d, the formation of mechanical twins inside austenite grains will promote the formations of successive mechanical twins by increased applied stress. The interactions of these mechanical twins form a cell structure, subdividing the austenite grains. As a strong obstacle against the movement of the dislocations, twin boundaries shorten the dislocation mean free path and promote dislocation pile-ups [34].

The strengthening contribution of the mechanical twinning can be modelled by the evolution of the dislocation density ρ with a shear strain of dislocation gliding γ_g as follows [31]:

$$\frac{d\rho}{d\gamma_g} = \frac{1}{b \cdot \Lambda} - f \cdot \rho \tag{3.2}$$

where b and f are the Burgers vector and the constant with a value of 3 for Fe-Mn-C steels, respectively. In Eq. (3.2), whilst the first term describes the storage of dislocations respecting the mean free path, the second term expresses the dislocation annihilation due to the dynamic recovery. The impact of the grain size d, average twin spacing t, and dislocation density ρ is expressed by the following equation [31]:

$$\frac{1}{\Lambda} = \frac{1}{d} + \frac{1}{t} + k \cdot \sqrt{\rho} \tag{3.3}$$

where k is a constant possesses a value of 0.011 for Fe-Mn-C steels. Finally, the average twin spacing can be estimated by a simple equation [31]:

$$t = 2e \cdot \frac{1 - F}{F} \tag{3.4}$$

where F is the volume fraction of the twins, e is the average twin thickness.

3.2.3 *Microband-Induced Plasticity (MBIP)*

The concept of microband was proposed for description of sheet like dislocation structure by Jackson [35]. Microbands consisting of geometrically necessary dislocations lead to the high dislocation density during deformation, resulting in continuous strain hardening by massive planar glide. These microbands formed by planar glide can accommodate a large amount of deformation [36], contributing to the development of high-performance MBIP steels. The SFE for MBIP steels is high and cross-slip of the screw component of partial dislocation is easy to occur, responsible for the formation of the geometrically necessary boundaries. Thus the term microband evolves as deformation mechanism [37]. Same as other microstructural features, *e.g.* boundary, cell structure, mechanical twins, microbands also impede the dislocation gilding effectively, leading to enhanced strain hardening capacity.

Park et al. [38] reported the deformation mechanism of in Fe-28Mn-12Al-1C (in wt.%) TRIPLEX steel [38]. Figure 3.5 shows the influence on deformation degree on dislocation planar glide [38]. Figure 3.5a illustrates the substructure for initial strain of ε = 5%. Dislocation segments move parallel aligned to the slip direction (planar glide). For strain at 10%, no uniform dislocation array, but the so-called Taylor lattice occurs Fig. 3.5b. This structure is arranged by distributed dislocations with planar ordered glide planes. These dislocations are quasi-uniformly ordered, resulting in an overall lower stored energy, caused by an alternating dislocation order. For increasing strain up to 40%, agglomerated geometrically necessary dislocations (GND) formed boundaries (Fig. 3.5c) [38]. The 60% strain micrograph obtains a regular appearing band structure, generated by GND. Disorientations on each side of the band are indicated by dashed lines (Fig. 3.5d). For 100% strain, the induced deformation leads to microstructural fraction due to intersecting microbands (Fig. 3.5e). The resultant domain boundaries and microbands subdivide the single austenite grain, as schematically shown in Fig. 3.5f [38]. The addition of Al in high-Mn steels promotes the precipitation of nanosized ordered $L'1_2$ $(Fe, Mn)_3AlC$ kappa phase. Gutierrez-Urrutia et al. [39] and Wu et al. [40] suggested that the interactions between microbands, dislocation structures and kappa phase accounted for the multiple stage strain hardening behaviours in MBIP steels.

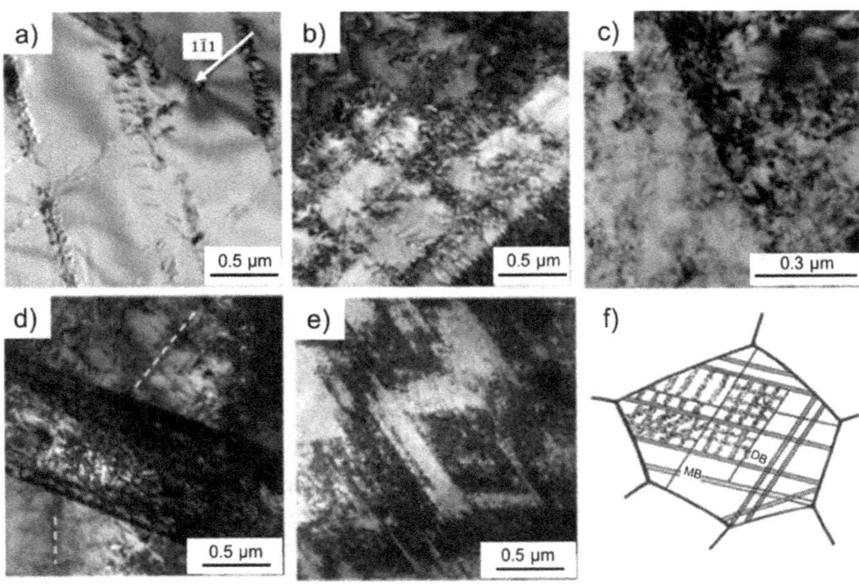

Fig. 3.5 TEM images of Fe-28Mn-10Al-1C alloy in varied nominal strain conditions: **a** Planar glide of dislocations ε = 5%, **b** Taylor lattice ε = 10%, **c** Taylor lattice domain boundary ε = 40%, **d** Microband ε = 60%, **e** Microband intersection with fraction ε = 100%, **f** Schematic of grain subdivision through microband intersection [38]

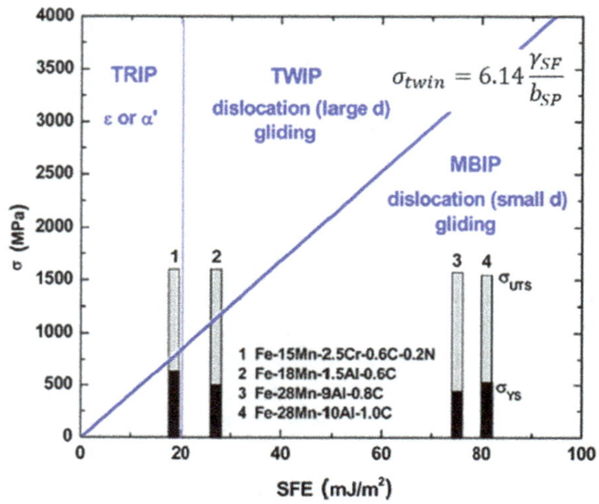

Fig. 3.6 Prediction of TWIP and MBIP transition in term of σ_{twin} in equation $\sigma_{twin} = 6.14\frac{\gamma_{SF}}{b_{SP}}$, for several TWIP and MBIP steels. The thin blue line indicates the boundary condition (20 mJ/m^2), from which the TWIP effect predominates the plastic deformation. The thick blue line denotes the critical applied stress to trigger the TWIP effect [41]

Figure 3.6 illustrates the relationship between SFE and the transition of deformation mode [41]. Microbands occur at the lower strain, benefiting from the enhancement of planar slip by shearable precipitates in the material [41]. The formation of nano-sized kappa phases (~10 nm) coherent to the fcc matrix promotes the formation of pronounced microband substructure [41]. Regardless of high SFE in the metallic materials, high solid solute atoms also promote planar glide to form Taylor lattice since solute interactions with dislocations increase the friction stress [41].

3.2.4 Dynamic Strain Aging (DSA)

The dynamic strain aging (DSA) effect may occur when the dislocation slip is slow enough to ensure the necessary time for the diffusion of atoms into the dislocation core that gives rise to the attachment and arrestment of a dislocation [42]. Several DSA mechanisms models were proposed. Lee et al. [43] proposed that DSA occurs by a single diffusive hop of the C atom in the point defect complex in the stacking fault region. The van den Beukel [44] suggested that solute clustering took place at temporarily arrested dislocations. The pinning force exerted by the solute clusters is a function of their size and depends on the solute diffusivity and the waiting time of the dislocation [44]. Mulford and Kocks [45] reported that the pipe diffusion of the solutes along the dislocation core may result in a high diffusivity without further needs of extra vacancies. Picu et al. [46] suggested that a pre-existing inhomogeneous solute distribution in the materials may cause DSA and influence the strain rate sensitivity. Rose and Glover [47] suggested that the presence of interstitial atoms, *e.g.* carbon, was essential in order to observe DSA in austenitic fcc alloys and the DSA effect was attributed to the reorientation of C-vacancy complexes in the stress field of dislocations [47].

3.3 SFE and 3D Deformation Mechanisms Map

3.3.1 Stacking Fault and Stacking Fault Energy

Austenite in steels is in fcc lattice structure with the close packed plane {111} and stacking in <111> direction with an ordered sequence of ABCABC. The change of stacking sequence is also possible, since the A, B, C positions for {111} plane are all energetically stable. Once atoms fall into a false position, the perfect stacking sequence may change into ABCABCACABCABC, leading to the formation of a stacking fault (SF), also known as intrinsic stacking fault [48]. ABCABCACB-CABCA type of plane stacking sequence may also occur where before and after C stacking sequence are both perfect types, also known as extrinsic stacking fault [48]. The most possible slip system in fcc crystal {111} <110> slip systems indicate the closest atom movement in close-packed plane [49]. As it is plotted in Fig. 3.7a, different colours represent consecutive close-packed planes ABC. A complete dislocation movement from position B to B needs to conquer an energy barrier from layer A [50]. Like it is easy to go across a mountain from the side than from the top, partials are formed to move from position B to C and then back to B position. In this way a complete $\frac{1}{2}$ [110] dislocation dissociates into two $\frac{1}{6}$ [211] partials [50]. Figure 3.7b presents the process of dislocation dissociation. The atoms between two partials, which are jumped from B position to C position as mentioned before, form a stacking fault [51]. When an extended dislocation goes across a sample during deformation, the first partial changes stacking sequence and the second restores it. The combined effect equals to the Burgers vector of a complete dislocation. SF is more frequently observed after deformation [52]. It could exist in a form of single layer, for example two partial dislocations, or cooperation of several layers, for example twinning, which could be understand as consecutive SFs at every atomic layer [48].

3.3.2 Experimental and Calculation Methods to Determine Stacking Fault Energy

3.3.2.1 Ab Initio Calculation

Ab initio methods have been applied recently to calculate the SFE in the materials [53–55]. Vitos et al. introduced a quantum mechanical first-principles approach, where the electronic, magnetic and volume effects were identified to be responsible for the compositional dependent SFE and the SF is a break of the ordered stacking sequence in {111} atomic layer [53]. In ab initio calculation, the SFE γ_{SF} is defined as follows [53]:

$$\gamma_{SF} = \frac{F_{SF} - F_0}{A_{2D}} \tag{3.5}$$

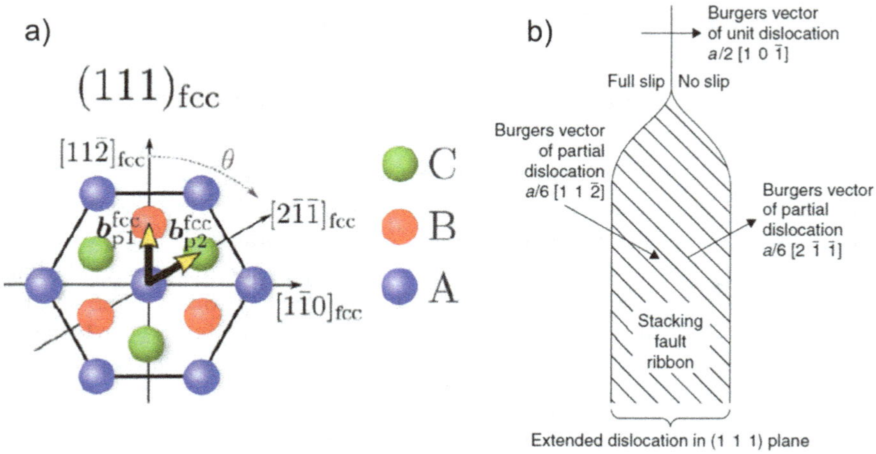

Fig. 3.7 **a** Schematic representation of crystal structure in fcc and the orientation relationship between complete and partial dislocations [50]; **b** Plotting of the process of SF formation by disassociation of complete dislocation into partials [51]

where F_{SF} and F_0 are the free energies of the system with and without the stacking fault, and A_{2D} denotes the area of the stacking fault.

Since the intrinsic stacking fault creates a negligible stress near the fault core, the faulted lattice almost preserves the close packing of the atoms, which can be modelled by an ideal close-packed lattice. Within the axial interaction model [56–59], the stacking sequence along the <111> direction is represented by a set of variables S_i that is determined by the translation connecting subsequent close-packed layers. The excess free energy of a particular stacking sequence is expanded as $-\Sigma_i \Sigma_j J_n S_i S_{i+n}$, where the sums run on the atomic layers, and J_1, J_2, ... are the nearest-neighbour, next nearest-neighbour, etc., interaction parameters. Using this representation, the energy of an intrinsic SFE can be expressed as

$$F_{SF} - F_0 = -4J_1 - 4J_2 - 4J_3 - O(J_4) \tag{3.6}$$

where $O(J_4)$ is the contribution arising from the higher order terms (J_4, J_5, ...).

3.3.2.2 Calphad

The SFs are treated either as volume Gibbs free energy difference between γ and ε [60, 61], as relating to interfacial energy [62, 63], or as comprising both volume and interfacial contribution [64, 65], in the thermodynamic assessments of SFE.

Ferreira and Müllner [66] suggested that SFs were embryo of martensite, and the SF was treated as a martensite with two atomic layers thickness in the calculation, where SFE is equal to the Gibbs free energy of a two atomic layer martensite formation. Specifically, the energy needed for a two atomic layer martensite formation is the summation of the volume Gibbs free energy difference between martensite and austenite matrix and the interfacial energy of upper and lower interface [66]. The equation is expressed as follows:

$$\gamma_{SF} = 2\rho \Delta G_{Eff}^{\gamma \to \varepsilon} + 2\sigma^{\gamma/\varepsilon} \tag{3.7}$$

where γ_{SF} denotes stacking fault energy, ρ is the molar surface density along {111} planes, $\Delta G_{Eff}^{\gamma \to \varepsilon}$ indicates Gibbs free energy due to the transformation of fcc austenite to hcp martensite, and $\sigma^{\gamma/\varepsilon}$ denotes the surface energy of austenite and martensite interface [66].

3.3.2.3 Transmission Electron Microscopy

A lot of efforts were made to determine the SFE by direct measurements employing transmission electron microscopy (TEM). Brown and Thölén [67] introduced an SFE evaluation by TEM observed dislocation structure. It is pointed out that even though, the relationship between the stacking fault energy and the radius of curvature of the partials in the node depended on whether the partials are edge-type or screw-type, it is possible to determine the stacking fault energy approximately by measuring the width of the node [67].

Figure 3.8 is a simplification of an extended three-fold node schema [67]. The SFE can be calculated according to following equation:

$$\gamma_{SF}/b = \sigma_{\infty}^{s} \tag{3.8}$$

where γ_{SF} denotes stacking fault energy and b is Burgers vector. Jandová [68] and Petrov [69] further measured SFE in Fe-Mn-Cr systems by performing the bright-field weak beam TEM technique to determine three-fold extended dislocation nodes

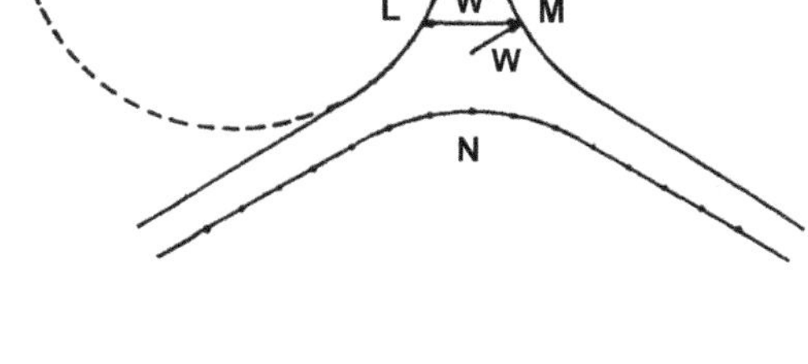

Fig. 3.8 A schema of an extended three-fold node used to evaluate SFE [67]

3.3.2.4 X-ray Diffraction

For the SFE measurement, X-ray method is statistically more meaningful compared with other mechanical and physical measurements. It was reported by Otte [70] that the stacking fault energy is proportional to the mean square strain and inverse proportional to the stacking fault density. For the mean square strain, it is proportional to dislocation density and can be obtained from measurement of the line broadening. In Adler's study [71], it was found that the shifts of the peak maxima are related to residual or applied stress σ, stacking fault probability α and redistribution of alloying elements, while the peak broadening results from the microstrain ε, and the stacking fault probability α as well. The ratio of the dislocation density ρ and the stacking fault probability are significantly related to the stacking fault energy. Reed and Schramm [72, 73] further developed the SFE determination by X-ray diffraction. From the microstrain and the stacking fault probability, the stacking fault energy can be determined:

$$\gamma_{SF} = \frac{K_{111}\omega_0 G_{(111)} a_0 A^{-0.37}}{\pi\sqrt{3}} \frac{\langle \varepsilon_{50}^2 \rangle_{111}}{\alpha} \tag{3.9}$$

where $K_{111}\omega_0$ is proportionality constant; $G_{(111)}$ is shear modulus in the (111) fault plane; $\frac{1}{3}(C_{44} + C_{11} - C_{12})$; a_0 is unit cell edge dimension; A is Zener anisotropy = $2C_{44}/(C_{11} - C_{12})$; ε_{50}^2 is root mean square (rms) strain, averaged over a column 50 Å; α is stacking fault probability and C_{ij} is elastic stiffness coefficients.

3.3.3 Thermodynamic Description of the Steels in Fe-Mn-C and Fe-Mn-Al-C Systems

Subregular model is a conventional method for thermodynamic calculation of SFE. Yang and Wan [74] introduced an SFE calculation model in Fe-Mn-Al-C system and indicated that the effective Gibbs free energy is summed by two parts. One is the chemical part, due to the chemical composition and lattice structure difference between γ and ε. The other one is contributed by transition of the paramagnetic to anti-ferromagnetic during $\gamma \rightarrow \varepsilon$ transformation [74]. The equation is presented in Eq. (3.10):

$$\Delta G_{Eff}^{\gamma\rightarrow\varepsilon} = \Delta G_{Chem}^{\gamma\rightarrow\varepsilon} + \Delta G_{Mag}^{\gamma\rightarrow\varepsilon} \tag{3.10}$$

$\Delta G_{Chem}^{\gamma\rightarrow\varepsilon}$ is the summation of individual influence of alloying elements on the total change of Gibbs free energy during $\gamma \rightarrow \varepsilon$ transformation. $\Delta G_{Mag}^{\gamma\rightarrow\varepsilon}$ is the magnetic contribution.

Regarding the chemical contribution, ideal entropy of mixing, which assumes the elements of a system to be a random mixing, is utilized. In this model, the interstitial elements like carbon have been treated as similar to Al and Mn in a substitutional solution [74]. The chemical contribution is calculated as follows:

$$\Delta G_{Chem}^{\gamma\rightarrow\varepsilon} = [\sum X_{i} \Delta G_{i}^{\gamma\rightarrow\varepsilon} + \sum X_i X_j \Delta\Omega_{ij}^{\gamma\rightarrow\varepsilon}] \tag{3.11}$$

where X_i and $\Delta G_i^{\gamma\rightarrow\varepsilon}$ indicate the molar fraction and the free energy change due to $\gamma \rightarrow \varepsilon$ transformation in pure metals. $\Delta\Omega_{ij}^{\gamma\rightarrow\varepsilon}$ denotes the binary interactions of elements.

Within the subregular model, the ternary interactions of elements are believed to be negligible, and Mn-Al, Al-C influences are small in iron rich material. Therefore, Eq. (3.11) is transferred into following in Fe-Mn-Al-C system steels [74]:

$$\begin{aligned}\Delta G_{Chem}^{\gamma\rightarrow\varepsilon} = & X_{Fe}\Delta G_{Fe}^{\gamma\rightarrow\varepsilon} + X_{Mn}\Delta G_{Mn}^{\gamma\rightarrow\varepsilon} + X_{Al}\Delta G_{Al}^{\gamma\rightarrow\varepsilon} + X_{C}\Delta G_{C}^{\gamma\rightarrow\varepsilon} \\ & + X_{Fe}X_{Mn}\Delta\Omega_{FeMn}^{\gamma\rightarrow\varepsilon} + X_{Fe}X_{Al}\Delta\Omega_{FeAl}^{\gamma\rightarrow\varepsilon} \\ & + X_{Fe}X_{C}\Delta\Omega_{FeC}^{\gamma\rightarrow\varepsilon} + X_{Mn}X_{C}\Delta\Omega_{MnC}^{\gamma\rightarrow\varepsilon}\end{aligned} \tag{3.12}$$

$\Delta G_{Mag}^{\gamma\rightarrow\varepsilon}$ as the magnetic parte is calculated by Eqs. (3.13) and (3.14):

$$\Delta G_{Mag}^{\gamma\rightarrow\varepsilon} = G_{Mag}^{\varepsilon} - G_{Mag}^{\gamma} \tag{3.13}$$

$$G_{Mag}^{\varphi} = RTln\left(1 + \frac{\beta^{\varphi}}{\mu B}\right) f\left(\frac{T}{T_{Neel}^{\varphi}}\right), \varphi = \gamma, \varepsilon \tag{3.14}$$

where G_{Mag}^{φ} is magnetic energy of fcc matrix or ε martensite, and β^{φ}, T_{Neel}^{φ}, μB indicate magnetic moment, Néel temperature of phase φ, Bohr magneton respectively. The published databases from Dew-Hughes and Kaufman [75], and Scientific Group Thermodata Europe (SGTE) [76] support the thermodynamic calculation of SFE.

3.3.4 *Thermodynamics-Based Deformation Mechanisms Maps Calculation*

The properties of high-Mn austenitic steels are strongly influenced by the SFE, which corroborates a specific deformation mechanism. Stacking fault energy is defined as the energy difference per unit fault area between faulty and perfect structures. The SFE controls the equilibrium distance between partial dislocations in fcc metals and hence the probability of activation of the various deformation mechanisms [77]. With the increasing SFE, the plasticity in the materials is achieved by dislocation glide and martensitic transformation (ε or α' martensite), dislocation glide and mechanical twinning, and pure dislocation glide [77].

The crucial requirement for the activation of supportive deformation mechanisms for dislocation slip, such as deformation-induced twinning and martensite formation, is a low SFE value [77]. Both TWIP and TRIP mechanisms are related to the dissociation of perfect dislocations into Shockley partials and thus are related to the energy of the created stacking faults. An SFE value of about 20 mJ/m^2 is considered to form the borderline between the TRIP and TWIP effect [78–82]. Therefore, the active plasticity mode is directly related to the SFE. The SFE can be calculated by thermodynamic modelling and be derived from ab initio calculations, as the decisive atomistic parameter [77]. These calculations are based on the Schrödinger equation for the binding energy between atoms and are therefore based on fundamental natural constants. Figure 3.9 shows the 2D composition-dependent SFE maps of high-Mn steels in Fe-Mn-C and Fe-Mn-Al-C alloy system [77]. The calculation is based on a subregular solution thermodynamics model. The maps are developed using the SGTE database [76]. The central assumption in the thermodynamic calculation is to take the stacking fault energy as the Gibbs free energy required to form a martensite plate two atomic layers thick within the dense planes [83, 84].

The corresponding deformation mechanisms of high-Mn steels, i.e. TRIP, TWIP, MBIP, are predicted in SFE maps based on the thermodynamic calculation. Figure 3.10 illustrates the different compositional ranges in terms of the different deformation mechanisms [77], where the kappa phase formation is calculated based on a Calphad approach. As shown in the composition-dependent 3D SFE map, the deformation mechanism shifts from TRIP to TWIP and on to MBIP with increasing SFE. With the extra addition of Al, kappa phase $(Fe, Mn)_3AlC$ forms in the Fe-Mn-Al-C steels according to the Calphad prediction. The planar slip accompanies the kappa phase formation with a significantly greater SFE value than the TWIP and TRIP effects.

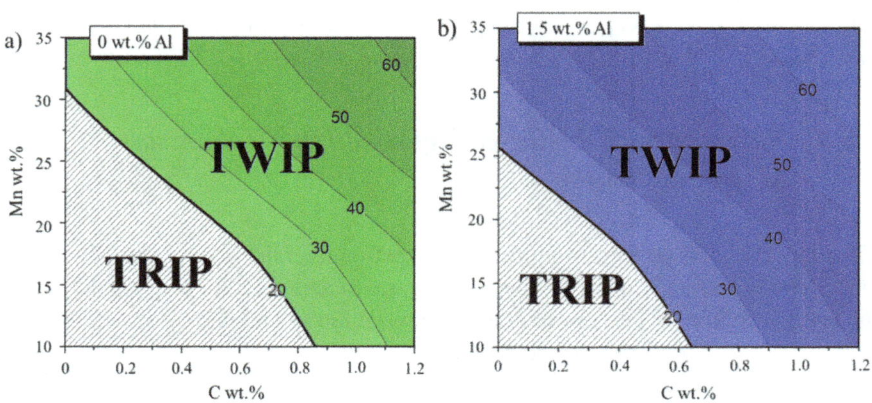

Fig. 3.9 2D composition-dependent map of stacking fault energy (unit, mJ/m^2) at 300 K. **a** Fe-Mn-C alloy system, **b** Fe-Mn-Al-C alloy system, Al 1.5 wt.% [77]

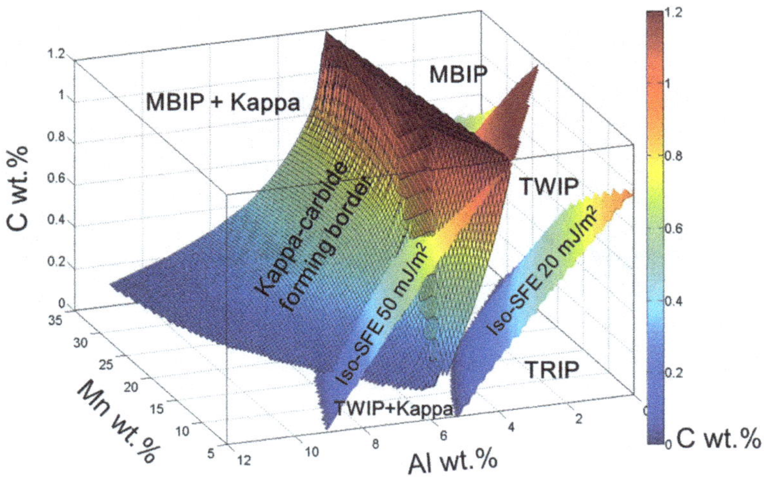

Fig. 3.10 3D composition-dependent SFE map of Fe-Mn-Al-C steel at 300 K [77]

3.4 Prediction and Control of Strain Hardening Behaviour in High-Mn Austenitic Steels

In this work, the composition-dependent SFE maps of high-Mn austenitic steels were calculated based on a subregular solution thermodynamics model [77]. The computed maps show the influence of chemical composition and temperature on the SFE and thus on the deformation mechanism. The strain hardening behaviour of austenitic steels with a high Mn content is described and elucidated using three application examples. The calculated mechanism maps can be used to facilitate material design to achieve the desired properties [77]. Additional features beyond explaining the SFE contribution to strain hardening behaviour will be further as follow.

3.4.1 Prediction of the Effect of Chemical Composition and Temperature on Mechanism Maps

The 3D plot in Fig. 3.11 shows the change of SFE with changing Al and C compositions in the composition-dependent SFE maps [77]. Figure 3.11a shows the SF maps with the Al content of 3, 5, 8 and 11 wt.%. The SFE is significantly increased by increasing the Al content, especially in alloys with a higher proportion of Mn and C. The addition of Al not merely elevates the stacking fault energy but also promotes the precipitation of nano-sized kappa phase (Fe, Mn)$_3$AlC, which regulates the mechanical properties [77]. The high strength and ductility of the high-Al alloyed steel is attributed to the precipitation of a nano-sized shearable kappa phase and its contribution in the development of planar dislocation substructures under applied stress [85].

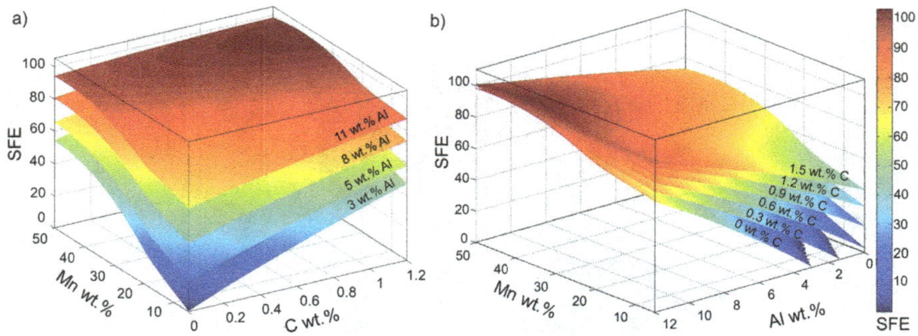

Fig. 3.11 **a** Effect of Al on composition-dependent SFE; **b** Effect of C on composition-dependent SFE [77]

As is demonstrated in Fig. 3.11b, the SFE also elevates with the increasing carbon contents [77]. The effect of carbon on SFE is more pronounced in the low Mn and low Al alloyed steels. While in the alloys with high Mn and high Al, the SFE is much elevated up to approximately 100 mJ/m^2 and the effect of C is not evident.

The 3D plot in Fig. 3.12a shows the iso-SFE surface at 20 mJ/m^2 (the approximate upper limit of TRIP mechanism [86]) and at 50 mJ/m^2 (the approximate upper limit of TWIP mechanism [41]). The parameters for calculating the temperature influence are taken from the literature [80, 87, 88]. The equations are used for calculation as follows [77]:

$$\Delta G_{Fe}^{\gamma \rightarrow \varepsilon} = -2243.38 + 4.309T \tag{3.15}$$

$$\Delta G_{Mn}^{\gamma \rightarrow \varepsilon} = -1000 + 1.123T \tag{3.16}$$

$$\Delta G_{Al}^{\gamma \rightarrow \varepsilon} = 2800 + 5T \tag{3.17}$$

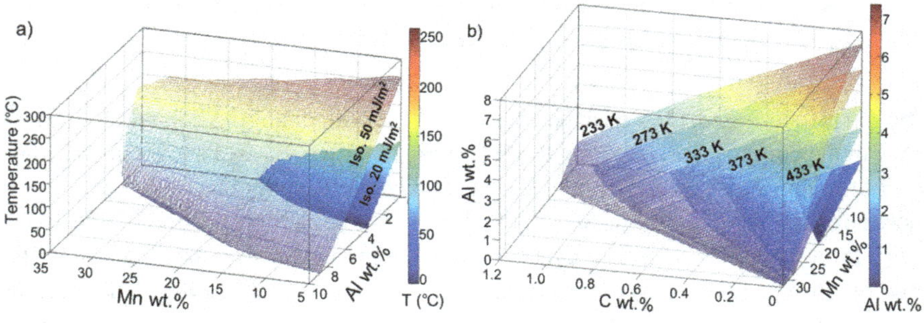

Fig. 3.12 Changing of iso-SFE surface with the increase of temperature [77]

$$\Delta G_C^{\gamma \to \varepsilon} = -22166 \tag{3.18}$$

The 3D plot in Fig. 3.12b shows the change of iso-SFE surface at 20 mJ/m^2 with temperature increase from 300 to 600 K in the composition-dependent SFE map [77]. The diagram illustrates the compositional domains in Fe-Mn-Al-C system steels. It also indicates that the stacking fault energy increases with increasing temperature. At higher temperature, the size of the dislocation nodes decreases and, for this reason, the stacking fault energy in the material can also increase [77].

The influence of temperature and chemical composition on the plastic deformation behaviour is shown by quasi-static tensile tests [77]. The strain was measured without contact using an optoelectronic video extensometer. A climate chamber was used for temperature variation tests. A thermocouple was applied directly to the sample surface to measure the temperature. The precision of the temperature control was in the range of ±2 °C. The strain rate for all tests was 0.001 s^{-1}. Samples with a gauge length of 30 mm were used for the tests and the elongation was calculated on the basis of a gauge length of 20 mm [77].

For the calculation of the strain hardening curve as the derivation $\theta = d\sigma/d\varepsilon$ of the flow curve $\sigma = f(\varepsilon)$, the data was previously smoothed by applying a weighted approximate non-parametric cubic spline to the flow curve [77]. With this method, oscillations and other discontinuities can be eliminated without impairing the characteristic course of the test. In addition, a light-optical analysis of the microstructure was carried out. The samples were electropolished to avoid mechanical manipulation of the surface. The microstructure was then revealed with Klemm etchant and examined [77].

Table 3.1 provides a summary of the materials tested along with the temperatures at which the tests were conducted, the grain sizes and the heat treatment [77]. The nominal chemical compositions of the four investigated materials are Fe-0.3C-17Mn-1.5Al (in wt. %), Fe-0.6C-17Mn-1.5Al (in wt. %), 0.3C-23Mn-1.5Al (in wt. %) and Fe-0.3C-28Mn (in wt. %). The amount of trace elements is <0.001% by weight in terms of phosphorus and <0.01% by weight in terms of sulphur. The choice of Al as an additional alloying element is based on its pronounced effect on SFE, making Al an excellent candidate for controlling the dominant deformation mechanism [89]. Moreover, Al has been reported to affect the onset of dynamic strain aging during plastic deformation [90]. The resulting mechanical properties of the investigated steels are summarized in Table 3.2.

3.4.2 Influence of Temperature on the Deformation Behaviour of Fe-0.3C-17Mn-1.5Al Steel

The influence of the test temperature on the deformation behaviour of a Fe-0.3C-17Mn-1.5Al (in wt. %) steel is shown in Fig. 3.13a for temperatures between −40 and 80 °C. The strength of the material increases with decreasing temperature [77].

Table 3.1 Testing temperature, grain size, SFE, annealing temperature and time, and the deformation mechanism of the investigated steels [77]

Steel	Test temperature (°C)	Grain size (μm)	Stacking fault energy (mJ/m^2)	Annealing temperature (°C)	Annealing time (min)	Deformation mechanism
0.3C-17Mn-1.5Al	-40	17.0	8.4	900	30	TRIP
	0	17.0	12.0	900	30	TRIP
	23	17.0	18.9	900	30	TRIP/TWIP
	80	17.0	26.9	900	30	TWIP
0.6C-17Mn-1.5Al	23	17.7	25.7	900	30	TWIP
0.3C-23Mn-1.5Al	23	17.9	27.5	900	30	TWIP
0.3C-28Mn	23	12.0	26.3	800	30	TWIP

Table 3.2 Mechanical properties of the investigated steels after recrystallization annealing [77]

Steel	Test temperature (°C)	Yield point (MPa)	Tensile strength (MPa)	Uniform elongation (%)	Total elongation (%)	Post uniform elongation (%)	ECO index UTS× El (GPa%)
0.3C-17Mn-1.5Al	−40	395	964	51.6	63.3	11.7	61.0
	0	321	808	53.6	59.2	5.6	47.8
	23	275	726	54.9	58.1	3.2	42.2
	80	220	609	64.0	71.2	7.2	43.4
0.6C-17Mn-1.5Al	23	290	751	57.2	60.6	3.4	45.5
0.3C-23Mn-1.5Al	23	222	612	59.1	64.7	5.6	39.6
0.3C-28Mn	23	294	745	52.3	53.9	1.6	40.1

According to the stress and strain engineering values in Table 3.2, the yield strength increases from 220 to 395 MPa and the tensile strength from 609 to 964 MPa. Concurrently, the total elongation is decreased from 71.2 to 63.3% [77]. At room temperature, the flow curve is almost linear. For 80 °C, the slope of the flow curve decreases, resulting in a convex shape and lower strength levels at high elongations. The course of the flow curves changes at test temperatures below room temperature. The work hardening rate decreases at low strains, but after a critical strain level it reaches an inflection point, the shape of the flow curve becomes concave [77]. This effect is more pronounced at −40 °C. A closer look at the deformation behaviour can be obtained by deriving the flow curves $\sigma = f(\varepsilon)$, these strain hardening curves $\theta = \mathrm{d}\sigma/\mathrm{d}\varepsilon = f(\varepsilon)$are plotted in Fig. 3.13b [77]. After a sharp drop at small strains, different behaviours can be observed depending on the temperature. From room temperature, the strain hardening rate decreases almost linearly with increasing elongation. At

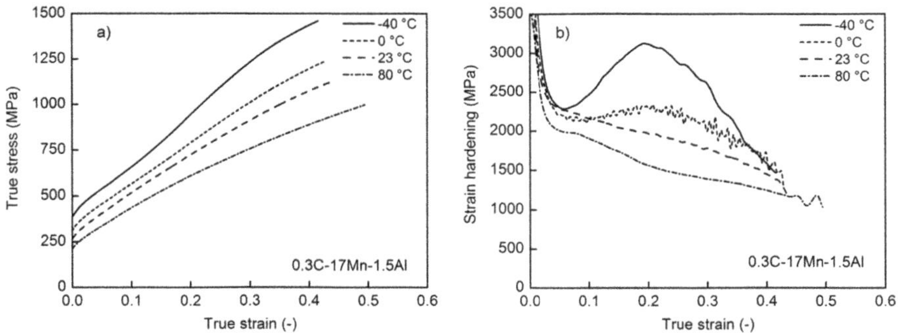

Fig. 3.13 **a** True stress strain and **b** Strain hardening curves of a Fe-0.3C-17Mn-1.5Al steel in a temperature range between −40 and 80 °C [77]

lower temperatures, θ increases and the formation of a local maximum can be seen which is more pronounced at −40 °C [77].

The results in Fig. 3.13 also show a change in strain mode as a function of temperature [77]. The predicted SFE at room temperature of Fe-0.3C-17Mn-1.5Al steel at 18.9 mJ/m^2 corresponds almost ideally to the expected transition between the TRIP and TWIP effects. A change in temperature affects the SFE and can therefore favour either TRIP at low temperatures or TWIP at temperatures above room temperature. Since the deformation mechanism has a direct influence on the resulting deformation structure, specimens deformed at −40, 23 and 80 °C were examined by light microscopy after fracture close to a crack and compared with the original structure before the test [77].

The undeformed Fe-0.3C-17Mn-1.5Al sample shows austenitic microstructure with an average grain size of 17 μm (Fig. 3.14a). After deformation at -40 °C, as shown in Fig. 3.14b, a high proportion of disordered structures can be observed within the grains [77]. After the room temperature deformation in Fig. 3.14c, parallel stripes within the grains become more visible. Finally, at 80 °C the grain boundaries of the initial microstructure are still visible. A relationship between strain hardening and deformed microstructure signifies the transition from TRIP to TWIP as predicted by the model. At −40 °C, the light optical microscopy (LOM) image shows typical martensitic microstructure, while at 80 °C, deformation twining can be observed. The mixed microstructure at room temperature shows the transition region where both deformation mechanisms occur simultaneously, which can be expected for a material close to the TRIP-TWIP transition line [77]. A quantity of 75 vol.% of martensite is found at −40 °C, which decreases to 23 vol.% at room temperature after shaping. At 80 °C, 5 vol.% of martensite is still present (Fig. 3.14d) [77].

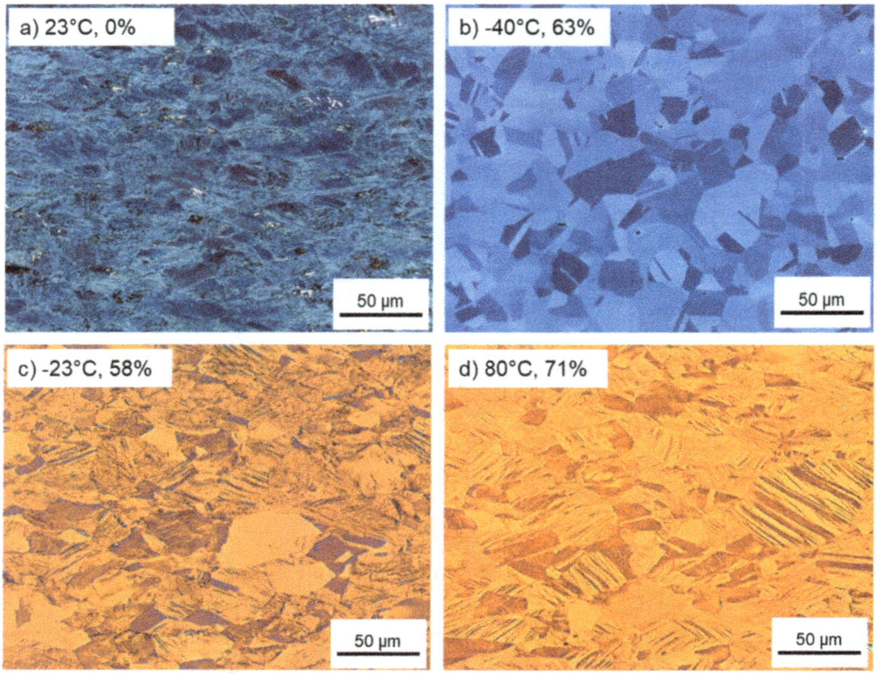

Fig. 3.14 Microstructure of a Fe-0.3C-17Mn-1.5Al steel in **a** initial state, and after fracture at **b** −40 °C, **c** −23 °C and **d** 80 °C (Klemm etchant, 500x magnification) [77]

3.4.3 Influence of Chemical Composition on Deformation Behaviours of Fe-Mn-C and Fe-Mn-Al-C Steels with iso-SFE

In this study, Al was added and the C and Mn contents were changed to obtain two alloys with the same SFE but significantly different compositions [77]. Figure 3.15 shows the flow curves in combination with the strain hardening curves of a Fe-0.6C-17Mn-1.5Al (in wt.%) steel with an SFE of 27.5 mJ/m^2 and a Fe-0.3C-23Mn-1.5Al (in wt.%) steel with almost the same SFE of 25.7 mJ/m^2. Referring to Table 3.1, both steels have similar initial grain size of 17.7 and 17.9 μm. It is found that the higher carbon steel has a noticeably higher yield strength. Concurrently, difference in tensile strength is even more profound. For plastic deformation, the steel with less carbon has a better post uniform elongation until fracture of 5.6% compared to 3.4%, while the uniform elongation is almost the same. After a small drop in the work hardening rate just after yielding, the flow curves are almost linear but with a different slope, resulting in a higher strength level of the Fe-0.6C-17Mn-1.5Al steel. The more pronounced strain hardening of the Fe-0.6C-17Mn-1.5Al steel is also evident in the strain hardening curves [77]. For both steels, the progression is almost parallel until a critical elongation of ~0.15 is reached. At higher strains, the

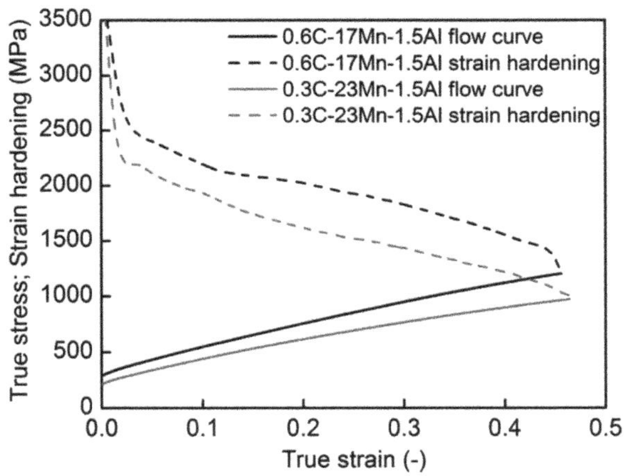

Fig. 3.15 Flow and strain hardening curves of a Fe-0.6C-17Mn-1.5Al and a 0.3C-23Mn-1.5Al steel [77]

slope of the Fe-0.6C-17Mn-1.5Al steel assumes a bulbous shape, while the work hardening rate of the Fe-0.3C-23Mn-1.5Al further decreases linearly. So, the higher carbon steel appears to have better hardening potential, even though both steels have the same SFE, resulting in mechanical twinning as the deformation mechanism [77].

Both microstructures in Fig. 3.16 show thin twins within the grains not crossing the grain boundaries. The number of these twin or slip bands is higher in the 0.6C-17Mn-1.5Al steel. Also, intersections of these bands can be observed in Fig. 3.16a, indicating the activation of a second twin system during deformation. Some of the straight twin bands evolve to a wavy form, which has also been previously observed [91] and explained by the interaction of the initial twins and slip [92]. As an interstitial alloy element, higher carbon contents have an increasing effect on the yield point. It was reported that the formation of the bulge in the strain hardening curve can be related with the tendency for twinning [93–95]. Carbon seems to be a key parameter for the onset of twinning, even in materials with similar SFE. It was reported that a Fe30Mn steel deformed only by slip mechanism while, for instance, a 0.6C-22Mn steel showed both, slip and twinning [93]. The same tendency can be concluded form the slope of the strain hardening curve in Fig. 3.15 and the corresponding microstructures are shown in Fig. 3.16 [77].

3.4.4 Influence of Chemical Composition on Strain Hardening of Fe-Mn-C and Fe-Mn-Al-C Steels with iso-SFE

The influence of the Al alloy on the work hardening behavior is investigated by comparing the steels Fe-0.3C-23Mn-1.5Al (in wt.%) and Fe-0.3C-28Mn (in wt.%) with SFE of 27.5 and 26.3 mJ/m^2, respectively [77]. Figure 3.17a shows the true

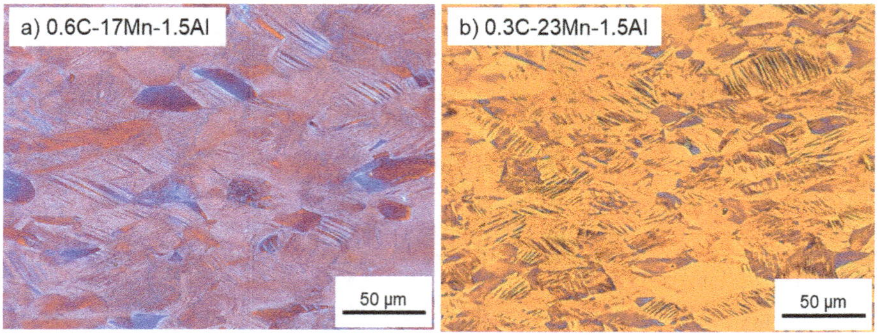

Fig. 3.16 **a** Microstructure of 0.6C-17Mn-1.5Al steel and **b** 0.3C-23Mn-1.5Al steel after fracture close to the crack; Klemm etchant, 500x magnification [77]

strain-stress curves and strain hardening behaviour of both steels. An obvious difference is that both steels have different grain sizes. With the size of 12 μm compared to 17 μm, the grains of the Fe-0.3C-28Mn are significantly smaller. Referring to Table 3.2, the Fe-0.3C-28Mn shows a higher yield strength and a higher tensile strength, better formability, indicated by the elongation values [77]. In particular, the uniform elongation at break after breakage is 3.5 times higher for Al-alloyed steel.

When comparing strain hardening rates, several differences can be observed. After yielding, the Fe-0.3C-23Mn-1.5Al shows a clear decrease with a linearly decreasing development up to uniform elongation. The Fe-0.3C-28Mn starts at a lower level, but the work hardening rate increases immediately and becomes concave after reaching a maximum at ~0.05 strain. After ~0.02 deformation, strain hardening is always higher compared to Fe-0.3C-23Mn-1.5Al steel. Figure 3.17b shows the results of the un-smoothened strain hardening rate curves for comparison [77]. As the load increases, both steels develop small spikes that are not regular in shape. From a degree of elongation of 0.3, the Fe-0.3C-28Mn shows clear peaks in the strain hardening curve with increasing maxima. After the third sharp peak, uniform strain is achieved, and

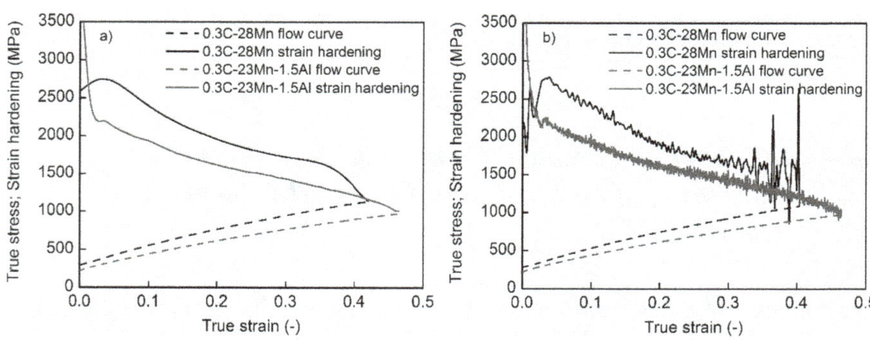

Fig. 3.17 **a** Smoothened and **b** Un-smoothened flow and strain hardening curves of a 0.3C-28Mn and a 0.3C-23Mn-1.5Al steels [77]

strain localization begins. These spikes can also explain the concave progression of strain hardening, as they have a positive effect on the overall hardening rate. However, the steel alloyed with Al does not show this effect [77].

The serrations shown in Fig. 3.17b are commonly reported in Fe-Mn-C steels and are described with the Portevin-Le-Chatelier (PLC) effect. The phenomenological background is not yet fully understood and is considered to be a dynamic interaction of dislocations or stacking faults with mobile solute atoms or point defect complexes [42, 43, 92]. The phenomenon is often referred to as DSA. The dislocations are blocked by the interaction with solute atoms, resulting in a higher critical stress for further motion. This discontinuous dislocation motion is often related to strain localization in bands propagating through the test specimen during tensile testing [77]. Several requirements must be met for the start of the DSA. At very low loads, high C contents can lead to the occurrence of DSA. For the Fe-0.6C-18Mn steel, the onset of DSA begins shortly after the plastic deformation [43]. For a material like the relatively low carbon Fe-0.3C-28Mn steel, a higher critical strain must be applied to generate the interaction between solute atoms and dislocation. This explains the late onset just before failure as shown in Fig. 3.17b. Due to the jagged flow behaviour in steels with DSA, early fracture is often observed because after a strain band has passed the material, the critical stress for the start of the next band exceeds the local UTS and leads to strain localization leading to fracture [77]. The Fe-0.3C-23Mn-1.5Al steel does not show DSA until fracture, resulting in a higher total elongation. The reason for the lack of DSA is identified as the addition of Al to suppress or retard DSA at higher loads or temperatures [43]. This effect can be clearly seen in the unsmoothed strain hardening curves in Fig. 3.17b and explains the improved post uniform elongation behaviour of the Fe-0.3C-23Mn-1.5Al [77].

3.5 Summary

The main factor that contributes to controlling the strain hardening behaviour of high Mn steels is the variants of SFE. The composition-dependent SFE maps calculated according to the subregular thermodynamic model can facilitate the prediction of deformation mechanisms in high-Mn steels in Fe-Mn-C and Fe-Mn-Al-C systems in 2D and 3D regimes [77]. However, a complete understanding of the control of work hardening behavior requires further comprehensive studies on specific characteristics, e.g. the effect of C, Mn and Al during dynamic strain aging (DSA) or their effect under a combined deformation mechanism of TWIP and DSA. Furthermore, Fig. 3.18 exhibits the potential of the presented engineering approach by comparing state of the art AHSS with a TWIP aided high-Mn steel (HMnS) and an austenitic CrNi stainless steel (CrNi) [77]. The nano-engineered steels can compete with other steels that have been developed according to the most modern material concepts, e.g. DP, low-alloy TRIP steels and austenitic CrNi stainless steels [77].

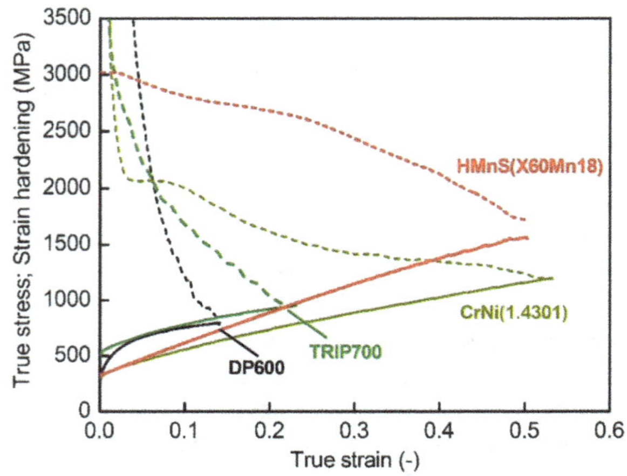

Fig. 3.18 Comparison of the flow and strain hardening behaviour of advanced high strength steels with TWIP aided and non-TWIP aided austenitic steels [77]

References

1. Bleck W, Brühl F, Ma Y, Sasse C (2019) Materials and processes for the third-generation advanced high strength steels. Berg Huettenmaenn Monatsh 164:466–474
2. Edmonds D, He K, Rizzo F, De Cooman B, Matlock D, Speer J (2006) Quenching and partitioning martensite—a novel steel heat treatment. Mater Sci Eng A 438–440:25–34
3. Tan X, Xu Y, Yang X, Wu D (2014) Microstructure-properties relationship in a one-step quenched and partitioned steel. Mater Sci Eng A 589:101–111
4. Shi J, Sun X, Wang M, Hui W, Dong H, Cao W (2010) Enhanced work-hardening behavior and mechanical properties in ultrafine-grained steels with large-fractioned metastable austenite. Scr Mater 63:815–818
5. Ma Y (2017) Medium-manganese steels processed by austenite-reverted transformation annealing for automotive applications. Mater Sci Technol 33:1713–1727
6. Bleck W, Guo X, Ma Y (2017) The TRIP effect and its application in cold formable sheet steels. Steel Res Int 88:1700218
7. Mahajan S, Green ML, Brasen D (1977) A model for the fcc → hcp transformation, its applications, and experimental evidence. Metall Trans A 8:283–293
8. Olson M (1972) A mechanism for the strain-induced nucleation of martensitic transformations. J Less Common Met 28
9. Glage A, Weidner A, Biermann H (2011) Cyclic deformation behaviour of three austenitic cast CrMnNi TRIP/TWIP steels with various Ni content. Steel Res Int 82:1040–1047
10. Hedayati A, Najafizadeh A, Kermanpur A, Forouzan F (2010) The effect of cold rolling regime on microstructure and mechanical properties of AISI 304L stainless steel. J Mater Process Technol 210:1017–1022
11. Vasilakos AN, Papamantellos K, Haidemenopoulos GN, Bleck W (1999) Experimental determination of the stability of retained austenite in low alloy TRIP steels. Steel Res Int 70:466–471
12. Souza Filho IR, Dutta A, Almeida Junior DR, Lu W, Sandim MJR, Ponge D, Sandim HRZ, Raabe D (2020) The impact of grain-scale strain localization on strain hardening of a high-Mn steel: real-time tracking of the transition from the $\gamma \rightarrow \varepsilon \rightarrow \alpha'$ transformation to twinning. Acta Mater 197:123–136
13. Lee S, Lee K, De Cooman BC (2015) Observation of the TWIP + TRIP plasticity-enhancement mechanism in Al-Added 6 Wt Pct medium Mn steel. Metall Mater Trans A 46:2356–2363
14. Li Z, Pradeep KG, Deng Y, Raabe D, Tasan CC (2016) Metastable high-entropy dual-phase alloys overcome the strength–ductility trade-off. Nature 534:227–230

15. Dan WJ, Zhang WG, Li SH, Lin ZQ (2007) A model for strain-induced martensitic transformation of TRIP steel with strain rate. Comput Mater Sci 40:101–107
16. Dan WJ, Li SH, Zhang WG, Lin ZQ (2008) The effect of strain-induced martensitic transformation on mechanical properties of TRIP steel. Mater Des 29:604–612
17. Oliver EC, Withers PJ, Daymond MR, Ueta S, Mori T (2002) Neutron-diffraction study of stress-induced martensitic transformation in TRIP steel. Appl Phys A 74:1143–1145
18. Abrassart F (1973) Stress-induced $\gamma \rightarrow \alpha$ martensitic transformation in two carbon stainless steels. Application to trip steels. Metall Trans A 4:2205–2216
19. Lo KH, Shek CH, Lai JKL (2009) Recent developments in stainless steels. Mater Sci Eng R Rep 65:39–104
20. Wong S, Madivala M, Prahl U, Roters F, Raabe D (2016) A crystal plasticity model for twinning- and transformation-induced plasticity. Acta Mater 118:140–151
21. Hong S, Han Y (1995) The effects of deformation twins and strain-induced ϵ-martensite on mechanical properties of an Fe-32Mn-12Cr-0.4C cryogenic alloy. Scr Metall Mater 32:1489–1494
22. Jacques P (2004) Transformation-induced plasticity for high strength formable steels. Curr Opin Solid State Mater Sci 8:259–265
23. Inegbenebor A, Jones R, Ralph B (1989) Mechanical properties and strain-induced phase transformations of some high strength manganese steels. J Mater Sci 24:3529–3535
24. Kim J, De Cooman BC (2016) Stacking fault energy and deformation mechanisms in Fe-xMn-0.6C-yAl TWIP steel. Mater Sci Eng A 676:216–231
25. Olson GB, Cohen M (1976) A general mechanism of martensitic nucleation: Part I. General concepts and the FCC $\rightarrow$ HCP transformation. Metall Trans A 12:1897–1904
26. Frommeyer G, Brüx U, Neumann P (2003) Supra-ductile and high strength manganese-TRIP/TWIP steels for high energy absorption purposes. ISIJ Int 43:438–446
27. Cohen J, Weertman J (1963) A dislocation model for twinning in f.c.c. metals. Acta Metall 11:996–998
28. Karaman I, Sehitoglu H, Gall K, Chumlyakov Y, Maier H (2000) Deformation of single crystal Hadfield steel by twinning and slip. Acta Mater 48:1345–1359
29. Mahajan S, Chin G (1973) Formation of deformation twins in f.c.c. crystals. Acta Metall 21:1353–1363
30. Steinmetz D, Jäpel T, Wietbrock B, Eisenlohr P, Gutierrez-Urrutia I, Saeed-Akbari A, Hickel T, Roters F, Raabe D (2013) Revealing the strain-hardening behavior of twinning-induced plasticity steels: theory, simulations, experiments. Acta Mater 61:494–510
31. Bouaziz O, Guelton N (2001) Modelling of TWIP effect on work-hardening. Mater Sci Eng. A 319:246–249
32. Idrissi H, Renard K, Ryelandt L, Schryvers D, Jacques PJ (2010) On the mechanism of twin formation in Fe-Mn-C TWIP steels. Acta Mater 58:2464–2476
33. Allain S, Chateau J, Dahmoun D, Bouaziz O (2004) Modeling of mechanical twinning in a high manganese content austenitic steel. Mater Sci Eng A 387:272–276
34. Chen L, Kim H, Kim S, De Cooman BC (2007) Localized deformation due to Portevin-LeChatelier effect in 18Mn–0.6C TWIP austenitic steel, ISIJ Int 47:1804–1812
35. Jackson PJ (1983) The formation of microbands by cross-slip. Scr Mater 17:199–202
36. Yoo JD, Hwang SW, Park KT (2009) Origin of extended tensile ductility of a Fe-28Mn-10Al-1C steel. Metall Mater Trans A 40:1520–1523
37. Huang JC, Gray G (1989) Microband formation in shock-loaded and quasi-statically deformed metals. Acta Metall 37:3335–3347
38. Park K (2013) Tensile deformation of low-density Fe-Mn-Al-C austenitic steels at ambient temperature. Scr Mater 68:375–379
39. Gutierrez-Urrutia I, Raabe D (2014) High strength and ductile low density austenitic FeMnAlC steels: simplex and alloys strengthened by nanoscale ordered carbides. Mater Sci Tech 30:1099–1104
40. Wu Z, Ding H, Li H, Huang M, Cao F (2013) Microstructural evolution and strain hardening behavior during plastic deformation of Fe-12Mn-8Al-0.8C steel. Mater Sci Eng A 584:150–155

41. Park K, Kim G, Kim S, Lee S, Hwang S, Lee C (2010) On the transitions of deformation modes of fully austenitic steels at room temperature. Met Mater Int 16:1–6
42. Kim J, Chen L, Kim H, Kim S, Estrin Y, De Cooman BC (2009) On the tensile behavior of high-manganese twinning-induced plasticity steel. Metall Mater Trans A 40:3147–3158
43. Lee S, Kim J, Kane S, De Cooman BC (2011) On the origin of dynamic strain aging in twinning-induced plasticity steels. Acta Mater 59:6809–6819
44. Van den Beukel A (1975) Theory of the effect of dynamic strain aging on mechanical properties. Phys Status Solidi A 30:197–206
45. Mulford R, Kocks U (1979) New observations on the mechanisms of dynamic strain aging and of jerky flow. Acta Metall 27:1125–1134
46. Picu R (2004) A mechanism for the negative strain-rate sensitivity of dilute solid solutions. Acta Mater 52:3447–3458
47. Rose K, Gloverj S (1966) A study of strain-ageing in austenite. Acta Metall 14:1505–1516
48. Christian J (2002) The theory of transformations in metals and alloys, 3rd edn. Elsevier, Oxford
49. Gottstein G (2004) Physical foundations of materials science, 1st edn. Springer, Oxford
50. Schönecker S, Li W, Vitos L, Li X (2021) Effect of strain on generalized stacking fault energies and plastic deformation modes in fcc-hcp polymorphic high-entropy alloys: a first-principles investigation. Phys Rev Mater 5:075004
51. Smallman R, Ngan A (2007) Physical metallurgy and advanced materials, 7th edn. Butterworth-Heinemann, Oxford
52. Reed-Hill R, Abbaschian R, Abbaschian L (2009) Physical metallurgy principles, 4th edn. Cengage Learning, Stamford
53. Vitos L, Nilsson J, Johansson B (2006) Alloying effects on the stacking fault energy in austenitic stainless steels from first-principles theory. Acta Mater 54:3821–3826
54. Dick A, Hickel T, Neugebauer J (2009) The effect of disorder on the concentration-dependence of stacking fault energies in Fe1-xMnx—a first principles study. Steel Res Int 80:603–608
55. Kibey S, Liu J, Curtis M, Johnson D, Sehitoglu H (2006) Effect of nitrogen on generalized stacking fault energy and stacking fault widths in high nitrogen steels. Acta Mater 54:2991–3001
56. Cheng C, Needs R, Heine V (1988) Inter-layer interactions and the origin of SiC polytypes. J Phys C: Solid State Phys 21:1049–1063
57. Denteneer P, van Haeringen W (1987) Stacking-fault energies in semiconductors from first-principles calculations. J Phys C: Solid State Phys 20:883–887
58. Denteneer P, Soler J (1991) Defect energetics in aluminium. J Phys: Condens Matter 3:8777–8792
59. Rosengaard N, Skriver H (1993) Calculated stacking-fault energies of elemental metals. Phys Rev B 47:12865–12873
60. Hendrickson A (1962) Solute segregation to stacking faults. Acta Metall 10:900–901
61. Flinn P (1985) Solute hardening of close-packed solid solutions. Acta Metall 6:631–635
62. Fine M, Guard R (1965) Surface thermodynamic treatment of adsorption on a dislocation-Suzuki locking. Trans Metall AIME 233:1383–1388
63. Hirth J (1970) Thermodynamics of stacking faults. Metall Trans 1:2367–2374
64. Olson G, Cohen M (1976) A general mechanism of martensitic nucleation: Part I. General concepts and the fcc $\rightarrow$ hcp transformation. Metall Trans A 7:1897–1904
65. Miodownik A (1978) The calculation of stacking fault energies in Fe-Ni-Cr alloys. Calphad 2:207–226
66. Ferreira P, Müllner P (1998) A thermodynamic model for the stacking-fault energy. Acta Mater 46:4479–4484
67. Brown L, Thölén A (1964) Shape of three-fold extended nodes. Discuss Faraday Soc 38:35–41
68. Jandová D, Řehoř J, Nový Z (2004) Microstructural changes taking place during the thermo-mechanical processing and cold working of steel 18Cr18Mn0.5N. J Mater Process Technol 157–158:523–530
69. Petrov Y (2005) On the electron structure of Mn-, Ni- and Cr-Ni-Mn austenite with different stacking fault energy. Scr Mater 53:1201–1206

70. Otte H (1967) Measurement of stacking-fault energies by X-ray diffraction. J Appl Phys 38:217–222
71. Adler R, Otte H, Wagner C (1970) Determination of dislocation density and stacking fault probability from X-ray powder pattern peak profile. Metall Trans 1:2375–2382
72. Reed R, Schramm R (1974) Relationship between stacking-fault energy and X-ray measurements of stacking-fault probability and microstrain. J Appl Phys 45:4705–4711
73. Schramm R, Reed R (1974) Stacking fault energies of seven commercial austenitic stainless steels. Metall Trans A 6:1345–1351
74. Yang W, Wan C (1990) The influence of aluminium content to the stacking fault energy in Fe-Mn-Al-C alloy system. J Mater Sci 25:1821–1823
75. Dew-Hughes D, Kaufman L (1979) Ternary phase diagrams of the manganese-titanium-iron and the aluminum-titanium-iron systems: a comparison of computer calculations with experiment. Calphad 3:175–203
76. Dinsdale A (1991) SGTE data for pure elements. Calphad 15:317–425
77. Song W, Ingendahl T, Bleck W (2014) Control of strain hardening behavior in high-Mn austenitic steels. Acta Metall Sin (Engl Lett) 27:546–556
78. Oh B, Cho S, Kim Y, Kim Y, Kim W, Hong S (1995) Effect of aluminium on deformation mode and mechanical properties of austenitic FeMnCrAlC alloys. Mater Sci Eng A 147–156
79. Allain S, Chateau JP, Dahmoun D, Bouaziz O (2004) Modeling of mechanical twinning in a high manganese content austenitic steel. Mater Sci Eng A 387–389:272–276
80. Bayraktar E, Khalid F, Levaillant C (2004) Deformation and fracture behaviour of high manganese austenitic steel. J Mater Process Tech 147:145–154
81. Renard K, Ryelandt S, Jacques P (2010) Characterisation of the Portevin-Le Châtelier effect affecting an austenitic TWIP steel based on digital image correlation. Mater Sci Eng A 527:2969–2977
82. Saeed-Akbari A, Imlau J, Prahl U, Bleck W (2009) Derivation and variation in composition-dependent staking energy based on subregular solution model in high-manganese steels. Metal Meter Trans A 40:3076–3090
83. Saeed-Akbari A, Mosecker L, Schwedt A, Bleck W (2012) Characterization and prediction of flow behavior in high-manganese twinning-induced plasticity steels: Part I. mechanism maps and work hardening behaviour. Metal Meter Trans A 43:1688–1704
84. Allain S, Chateau J, Bouaziz O (2004) A physical model of the twinning-induced plasticity effect in a high manganese austenitic steel. Mater Sci Eng A 387–389:143–147
85. Gutierrez-Urrutia I, Raabe D (2013) Influence of Al content and precipitation state on the mechanical behavior of austenitic high-Mn low-density steels. Scr Mater 68:343–347
86. Mosecker L, Saeed-Akbari A (2013) Nitrogen in chromium-manganese stainless steels: a review on the evaluation of stacking fault energy by computational thermodynamics. Sci Technol Adv Mater 14:033001
87. Saeed-Akbari A (2011) Mechanism maps, mechanical properties, and flow behaviour in hign-manganese TRIP/TWIP and TWIP steels. Doctoral thesis, RWTH Aachen University
88. Jiang H, Zhang Q, Chen X, Chen Z, Jiang Z, Wu X, Fan J (2007) Three types of Portevin-Le Chatelier effects: experiment and modelling. Acta Mater 55:2219–2228
89. Dumay A, Chateau J, Allain S, Migot S, Bouaziz O (2008) Influence of addition elements on the stacking-fault energy and mechanical properties of an austenitic Fe-Mn-C steel. Mater Sci Eng A 483:184–187
90. Park K, Jin K, Han S, Hwang S, Choi K, Lee C (2010) Stacking fault energy and plastic deformation of fully austenitic high manganese steels: effect of Al addition. Mater Sci Eng A 527:3651–3661
91. Bäumer A, Jiménez J, Bleck W (2010) Effect of temperature and strain rate on strain hardening and deformation mechanisms of high manganese austenitic steels. Int J Mater Res 101:705–714
92. Meng L, Yang P, Xie Q, Ding H, Tang Z (2007) Dependence of deformation twinning on grain orientation in compressed high manganese steels. Scr Mater 56:931–934
93. Bouaziz O, Allain S, Scott C, Cugy P, Barbier D (2011) High manganese austenitic twinning-induced plasticity steels: a review of the microstructure properties relationships. Curr Opin Solid State Mater Sci 15:141–168

94. Gutierrez-Urrutia I, Raabe D (2011) Dislocation and twin substructure evolution during strain hardening of an Fe-22 wt.% Mn-0.6 wt.% C TWIP steel observed by electron channeling contrast imaging. Acta Mater 59:6449–6462
95. Bracke L, Kestens L, Penning J (2009) Direct observation of the twinning mechanism in an austenitic Fe-Mn-C steel. Scr Mater 61:220–222

Chapter 4
Nanostructure Characterization Methods

This chapter provides the description of advanced techniques for nanostructure characterization, such as high energy synchrotron X-ray diffraction (SYXRD), small angle neutron scattering (SANS), atom probe tomography (APT), and high-resolution transmission electron microscopy (HRTEM). The fundamentals and development history of these material characterization methods are introduced. Besides, the experimental setups and sample preparation procedures are further addressed.

4.1 High Energy Synchrotron X-Ray Diffraction (SYXRD)

4.1.1 Principle of SYXRD

X-ray is a kind of high energy electromagnetic spectrum with a wavelength of an order smaller than nanometre at the similar level as the atom distances [1]. The development of the materials characterization with X-ray started 100 years ago since W.C. Roentgen discovered X-ray. X-rays generated in synchrotron storage rings are characterized by their high intensity, high energy, and parallel and narrow structure, i.e. the high brilliance and high flux [2]. Flux and brilliance respectively imply the number of photons per second at 0.1% bandwidth, and the flux per unit source area and unit solid angle. The higher brilliance of X-ray source makes possible to focus a beam onto a fine spot [3]. These specific features bring the advantage of in-situ characterization of centimetre thick specimens within a second. With a brightness of a million times greater than the sun and that of conventional X-ray tubes, the acquired signal from the emitted photons is high enough that samples with very dilute concentration of the investigated species can be measured.

Unlike standalone X-ray tubes, synchrotron beamlines are long straight beams, through which any tailor-made experimental stations can be installed. The pulsed

W. Song, *Nano-Engineering of High Strength Steels*,
Topics in Mining, Metallurgy and Materials Engineering,
https://doi.org/10.1007/978-3-031-42967-5_4

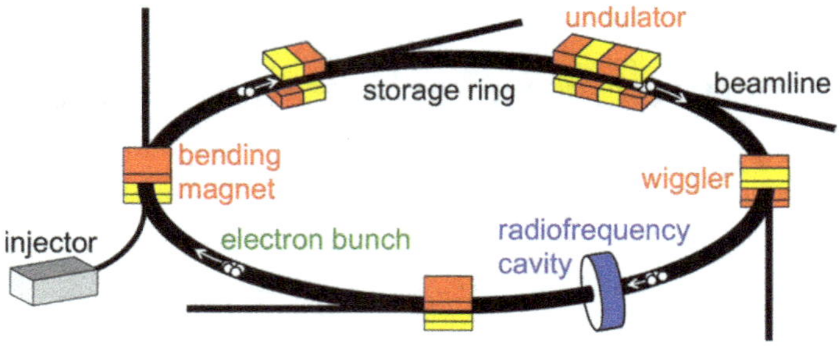

Fig. 4.1 General scheme of a synchrotron facility [5]

time structure due to bunches of produced electrons allows up to picoseconds of time-resolved investigation. Its defined polarization realizes the possibility to investigate magnetic materials. It possesses also a high degree of collimation and hence increases the energy resolution. In the synchrotron X-ray sources, electrons are forced by a magnetic field to move along curved trajectories. The movement of charged particles such as electrons at relativistic speeds, i.e. close to the speed of light, results in an emission of electromagnetic radiation in the direction of motion [4]. Figure 4.1 schematically represents the four main structural parts of a synchrotron facility, namely, (i) the electron injector, (ii) the storage ring (accelerator), (iii) the radiofrequency cavity, and (iv) X-ray sources of different types with their beamlines [5].

The materials microstructure composes of grains, phases and various defects, e.g. dislocations, stacking faults, grain boundaries and phase boundaries. Diffraction methods allow qualitatively and quantitatively characterizing the microstructural constituents. When a metallic material is exposed to the incident beam of the X-ray, neutron, or electron that possesses wavelength in the same order with the interplanar spacing, constructive or destructive interference may take place along different directions due to the coherent elastic scattering of crystal atoms [6]. Diffraction occurs when the coherent elastic scattering of planes is in the same phase and the path length difference between adjacent planes ($2d \sin \theta$) is equal to λ [6]. The diffraction of a crystal obeys the Bragg's law, shown in the following equation:

$$\lambda = 2d \sin \theta \tag{4.1}$$

where λ is the wavelength of an incident beam, d is the interplanar distance, and θ the angle of incidence and reflection. Figure 4.2 schematically illustrates the experimental setup of powder diffraction in a synchrotron facility [7, 8]. The detailed examination of diffraction peaks with regards to their intensity, position, broadening and anisotropy makes possible to characterize the microstructure, e.g. phase fraction, dislocation density, crystallite size and stacking fault probability [7, 8].

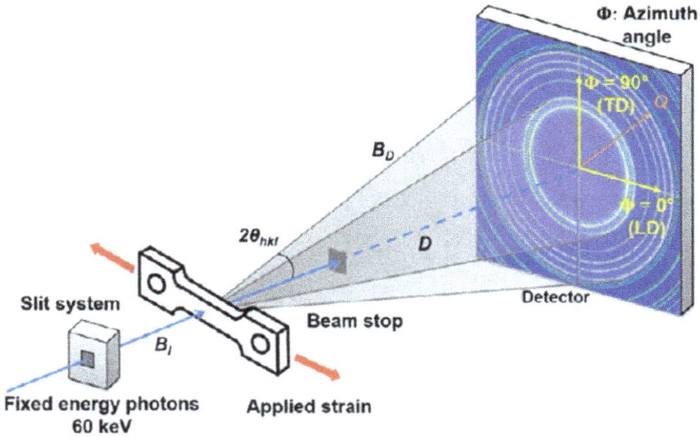

Fig. 4.2 Schematically representation of the experimental setup of powder diffraction in a synchrotron facility [8]

4.1.2 Instrumental Set-Up and Sample Preparation

4.1.2.1 Beamline Alignment

The alignment of high-resolution powder diffraction beamline P02.1 at the Deutsches Electron-Synchrotron centre (DESY) is introduced as an example in this sub-chapter. It is part of the Hard X-Ray Diffraction beamline P02. It operates at a fixed energy of 60 keV with the wavelength of 0.207 Å and exhibits low divergence, small energy bandwidth, and high flux [8, 9]. The photon energy of 60 keV is particularly useful for the characterization of nano-crystalline and disordered materials [8, 9]. Thus, total scattering experiments that evaluate data in real and reciprocal space are predestined for P02.1, and the high photon flux enables time-resolved studies with resolution down to the (sub) second regime [8, 9]. As a result, structural and chemical transformations in this time regime can be effectively studied [8, 9].

4.1.2.2 Ex-Situ Synchrotron X-Ray Diffraction Measurement

Figure 4.3a schematically illustrates the major structural components of P02.1 beamline, i.e. the undulator (U), slit systems (S1 and S2), diamond and silicon Laue monochromator crystals (L1 and L2), sample holder (SP), multi-analyser crystal (D1) and area detector (D2) [9]. Figure 4.3b shows the setup of the experimental hutch in which beam comes in from the right-hand side. The alignment of sample position is done by using laser guideline and the experiments are controlled in an operation computer. The sample to area detector distance is adjusted as 1 m. Thanks to the high flux of the beam, each diffraction pattern is able to be recorded within a period down to sub-second [8, 9].

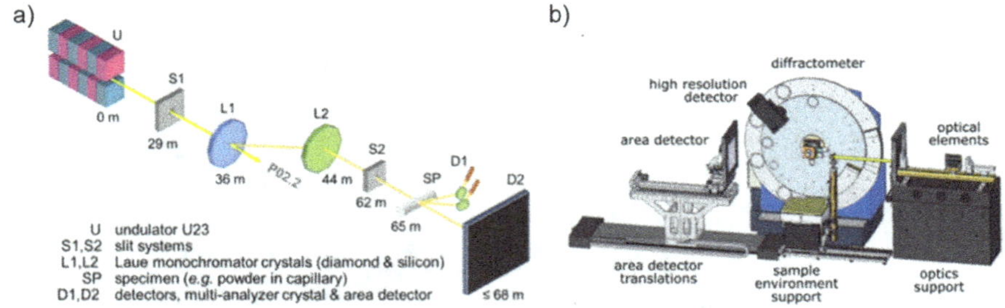

Fig. 4.3 **a** Schematic setup of P02.1 beamline with major structural elements and their distances to the undulator source; **b** Schematic setup of the experimental hutch in which beam comes in from the right-hand side [9]

4.1.2.3 In-Situ Heating and Deformation Measurement

Prior to the in-situ heat treatments of the materials at the beamline P02.1 at DESY, the samples are sealed in capillary to prevent oxidation during the annealing. The samples are prepared by precise cutting machine into dimensions of 10 × 1 × 1 mm, and then mechanically ground to a cross-section of dimensions down to ~0.8 × 0.8 mm. The samples are aligned at the centre of the furnace and X-rays can perpendicularly penetrate through the sample. During heat treatments of the materials, the diffraction patterns can be acquired with specific time resolution down to sub second to characterize the microstructural evolution at the beamline. During in-situ tensile test, the samples are set at the central of the tensile box and the X-rays perpendicularly pass through the sample in the thickness direction. The grip of the tensile box moves at a constant velocity of 20 μm/s. Owing to the in-situ method and the high intensity of synchrotron X-rays, the microstructure evolution during heat treatments and during deformation in the materials can be precisely captured.

4.1.3 Data Analysis Based on Rietveld Method

Rietveld refinement method is an effective approach to analyse the synchrotron xrd data, considering both structureal model and the experimental diffraction data. It assumes that the best condition of the refinement is when the value is the smallest in Newton–Raphson algorithm:

$$S_y = \sum_i W_i (Y_i - Y_{ci})^2 \tag{4.2}$$

where, Y_i is the observed intensity at a certain 2θ; Y_{ci} is the calculated intensity at the same angle; W_i is a weight, $W_i = 1/Y_i$; $i = 1, 2, \ldots n$; α is the parameters to be refined [10, 11]. Hill et al. [10] introduced a model based on Rietveld method to identify the phase fraction of a crystalline material, where the weight fraction (W) of a phase p is proportional to the Rietveld scale factor (S) and the weight fraction of a phase $p\left(W_p\right)$ can be calculated as follows [10]:

$$W_p = \frac{S_p(ZMV)_p}{\sum_i S_i(ZMV)_i} \tag{4.3}$$

where Z, M and V refer respectively as to the number of phase constituent per unit cell (e.g. an austenitic Fe unit cell contains 4 Fe atoms), the mass of phase constituent, and the volume of a unit cell.

Peak ratio method can also be considered as a phase fraction measurement technique for the material system possesses a mixture of maximum two phases [10]. The ratio between the phases (x_A/x_B) is calculated by the following equation [10]:

$$\frac{x_A}{x_B} = \left(\frac{8|F_B|^2 V_A^2}{12|F_A|^2 V_B^2} \cdot \frac{1+\cos^2\theta_B}{\sin\theta_B \cdot \sin 2\theta_B} \cdot \frac{\sin\theta_A \cdot \sin 2\theta_A}{1+\cos^2\theta_A}\right) \cdot \frac{I_A}{I_B} \tag{4.4}$$

where I_i is the integrated intensity of a diffraction peak; F_i is the structure factor of a unit cell that is equal to the multiplication of atom scattering factor and number of atoms per unit cell; θ_i and V_i imply the peak position and the volume of a unit cell, respectively. However, this method is limited in considering the absorption factors.

Furthermore, high energy synchrotron X-rays with high brilliance, small beam size and high penetration depth is helpful to study the dynamics of dislocation, stacking fault and twinning evolution during in-situ tensile test of high manganese Fe-Mn-Al-C alloys. The 2D diffraction patterns of Fe-17Mn-1.5Al-0.3C steel during deformation are shown in Fig. 4.4. The grain refinement is obviously seen from the clear increase of the intensity of the 2D diffraction rings.

The analysis of crystal structure parameters, phase fractions, extrinsic and intrinsic stacking fault probabilities, crystallite size and microstrain values can be conducted using the Rietveld method [11] based the Materials Analysis Using Diffraction (MAUD) software [12, 13]. The upper window in Fig. 4.5 shows the experimental (black line) and the simulated (red line) 1D diffraction patterns of the 30 min annealed and 25% tensile deformed X60Mn18 steel. The lower window represents the difference in the intensities of the exp&simu 1D diffraction patterns.

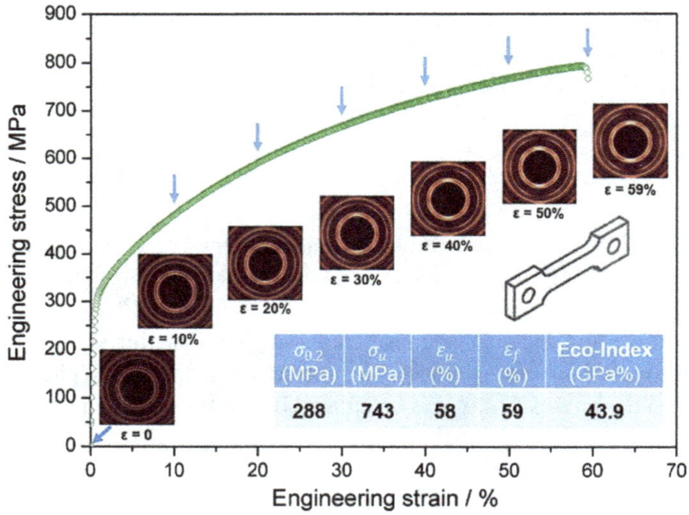

Fig. 4.4 2D diffraction patterns of Fe-17Mn-1.5Al-0.3C steel during deformation (strain rate: $0.001\ s^{-1}$)

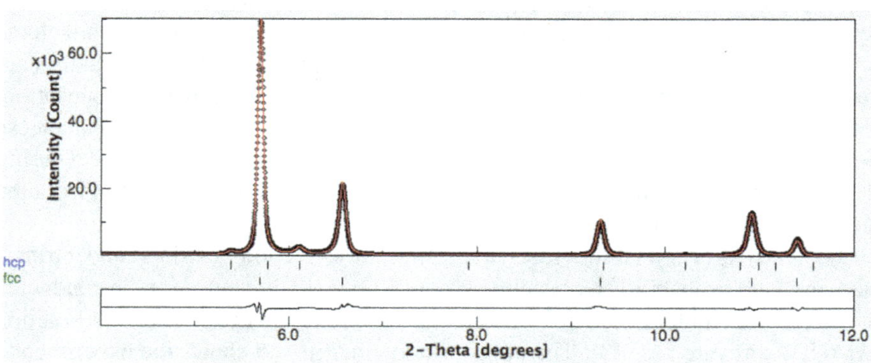

Fig. 4.5 Fitted diffraction profile of high-Mn TRIP/TWIP steel X60Mn18 with consideration of stacking fault and twinning models

4.2 Small Angle Neutron Scattering (SANS)

4.2.1 Introduction to SANS

4.2.1.1 Theory of SANS

SANS is a well-established characterization method for microstructure investigation in various materials. It can typically probe inhomogeneities in the nanometre scale (1–100 nm). Since the construction of the first SANS instruments in 1970s, this technique experienced a steady development over half a century [14]. SANS instruments are either based on the neutron reactors using monochromat neutron beams or time-of-flight (TOF) instruments at a pulsed neutron source. The aim of SANS experiments is to determine the shape and organization averaged in time of particles or aggregates dispersed in a continuous medium. In the following, emphasis is given to continuous reactors and spallation sources [15]. In order to obtain insights into the structure of matter, the used wavelength has to be on the atomic scale or rather smaller than the inter-distance spacing between the particles, and it allows detecting low volume fractions (up to 10^{-4}) thanks to its extraordinary high sensitivity [16, 17].

SANS is a non-destructive method and neutrons exhibit more penetrating than X-rays. While X-rays interact with the electron cloud of the atom, the neutron interact with the nuclei of the atom. Neutrons also shows the major advantage in interacting differently with hydrogen and deuterium. An often-mentioned drawback of neutrons is the low beam flux compared to X-ray sources, since it lowers the scattering intensity and thus requires longer measuring time [17]. The scattered neutrons facilitate the analysis of the materials structures based on not only the sharp areas of the sample, but the whole irradiated area throughout thicknesses up to a few cm [18, 19]. Therefore, SANS is statistically more representative than transmission electron microscopy (TEM), which probes precisely the local structures in the materials with a thickness in the range of a few nanometers [18, 19].

4.2.1.2 Scattering Parameters

As illustrated in Fig. 4.6, neutrons are scattered under an angle θ, where the detector is irradiated by a fraction of neutrons inside the solid angle dΩ and the scattering intensity is measured by a 2D detector in dependence of the scattering angle 2θ [20, 21]. The sample to detector distance (SDD) is suggested to be large compared to the detector area, hence the solid angle θ becomes small [20, 21]. The number of neutrons flowing on an area perpendicular to the neutron beam per second defines the flux ψ of a neutron beam [22]. The neutron scattering cross section Σ describes the effective area of the scattering body perpendicular to the incident neutron beam, which possesses a unit in area [22].

The following equation describes the relationship between the neutron scattering cross section $\sum$ and incident neutron beam flux ψ [22]:

$$\Sigma = \frac{1}{\psi} \cdot \text{number of neutrons scattered per second} \tag{4.5}$$

$\sum$ reflects the total cross section created by scattering body and depends on the volume of a sample. In order to attain sample specific, i.e. intrinsic property, neutron scattering cross section, $\sum$ is normalized by the volume of a sample [22]:

$$\sigma = \frac{1}{V} \cdot \Sigma \tag{4.6}$$

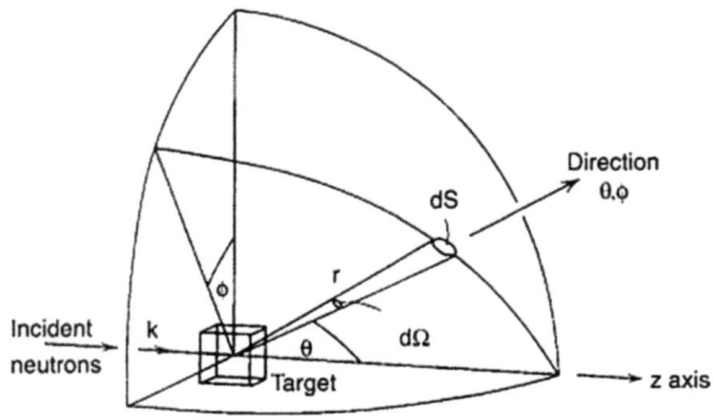

Fig. 4.6 Schematic illustration of the scattering geometry [21]

where σ is the volume specific cross section [22, 23]. The differential scattering cross section reflects the angular dependence of the scattered neutrons as follows:

$$\frac{d\sigma}{d\Omega} = \frac{1}{V} \cdot \frac{d\Sigma}{d\Omega} \tag{4.7}$$

with $d\Omega$ is the solid angle. The sum of the scattered neutrons in all of the solid angle, i.e. 4π, gives the total number of scattered neutrons.

$$\sum = \int \frac{d\sum}{d\Omega} d\Omega \tag{4.8}$$

The scattering flux, i.e. scattering intensity per second and solid angle, at a given neutron wavelength and the scattering vector $I(\lambda, q)$ is expressed as follows [22]:

$$I(\lambda, q) = \psi \cdot \Delta\Omega \cdot \eta(\lambda) \cdot T \cdot V \frac{d\sigma}{d\Omega}(q) \tag{4.9}$$

where ψ is the incident flux, $\Delta\Omega$ is the detector area regarding the solid angle, $\eta(\lambda)$ is the detector efficiency, T is the transmission value of a sample, V is the illuminated sample volume. The significant structural information about the scatters such as the size, shape, number density, volume and their interactions are extracted from the differential scattering cross section $d\sigma/d\Omega(q)$.

$$\frac{d\sigma}{d\Omega}(q) = (\Delta p)^2 \cdot \int_0^\infty N(R) \cdot V(R)^2 \cdot F(q, R)^2 \cdot S(q, R_{HS}, f_{HS}) dR \tag{4.10}$$

where Δp is the scattering length density difference of the matrix and scatter, $N(R) \cdot dR$ is the number-density size distribution of scatters with sizes between R and $R+dR$, $V(R)$ is the volume of a scatter, $F(q, R)$ is the shape factor, $S(q, R_{HS}, f_{HS})$ is the structure factor [24].

4.2.2 Experimental Set-Up and Sample Preparation

Figure 4.7 shows a schematic illustration of SANS experimental principle at a continuous neutron source [24]. Both the nuclear reactor and spallation source can be used for the neutron production [25, 26]. The neutron beam with a few millimetres of cross-section is required in order to cover small angles for the scattering and the focused small cross-sectioned neutron beam is obtained by collimation system [27–29]. The scattering pattern including the information on the scattering intensity at different scattering angles, is recorded by the 2D position sensitive detector.

Figure 4.8 shows the KWS-2 beamline at Heinz Maier-Leibnitz (MLZ) at Garching, Germany. The magnetic field is applied vertically and perpendicular to the incoming neutron beam. Sample to detector distance can be set to 2 m, 8 m and 16 m for each sample, respectively. The incoming neutron flux on the sample is typically around $2.5 \times 10^7 \cdot 1/\text{cm}^2\text{s}$ up to $2 \times 10^9 \cdot 1/\text{cm}^2\text{s}$. With different sample to detector distances, a wide q-range can be covered (1×10^{-4} 1/Å – 0.5 1/Å) [30].

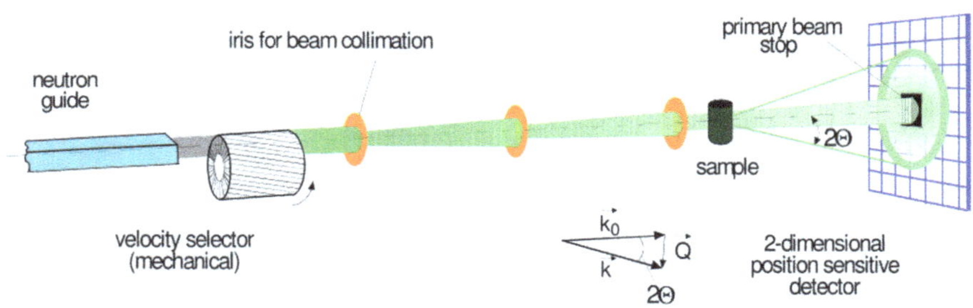

Fig. 4.7 Schematic illustration of a SANS experimental principle at a continuous neutron source [24]

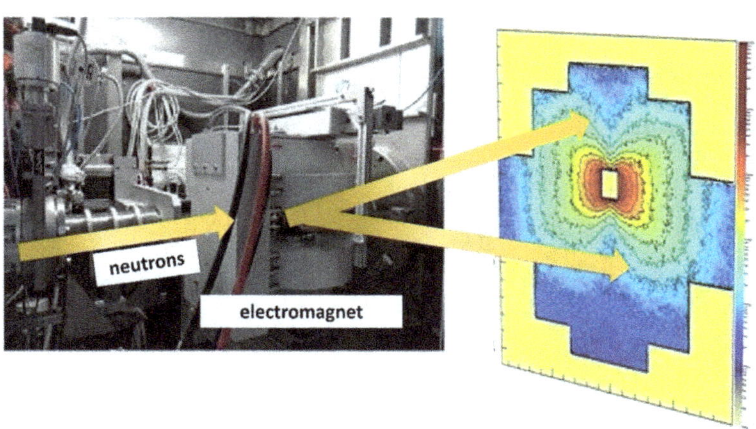

Fig. 4.8 KWS-2 beamline at Heinz Maier-Leibnitz (MLZ) at Garching, Germany

4.2.3 Data Analysis

In addition to transmission of a sample, the type, size and number distribution of the inhomogeneity (e.g. cluster, particle) content determine the scattering intensity of each scattering vector. Structural information of a sample can be obtained from the examination of the intensity. Figure 4.9 shows a one-dimensional scattering pattern of the material that is obtained by using different sample to detector distances.

Besides using the nonlinear least-squares fitting models, the SANS data can be analysed by using the slopes of the scattering curve [31–35]. The power law gives information about the nature of the scattering objects. If the decay of the scattering intensity follows a q^{-D} behavior, where D is the fractal dimension, i.e. Porod exponent, the initial information about the structure of the sample can be obtained. Different D values 1, 2 and 4, respectively, represent the elongated (rod-like), flattened (disc-like) and sharp interfaces [35]. In other words, whilst $D = 4$ reflects the scattering objects with smooth surfaces, $D = 3$ indicates every rough surfaces, $D = 2$ points the presence of the two-dimensional structures, e.g. lamellae or platelets and $D = 1$ reflects the scattering from thin cylinders [31]. The Guinier equation, which is valid for the small q values, describes the relation between intensity and q King [31].

$$I_{(q)} \approx I_{(0)} e^{-\left(\frac{1}{3}\right) q^2 R_g^2} \tag{4.11}$$

R_g is the "weight average" of all radii presented in the sample. It can be used to calculate average radius R_m of the particles (scatters) in the matrix. The relation between R_m and R_g is different for different scatter shapes. The radius of spheres can be calculated by:

$$R_g^2 = \frac{3}{5} R_m^2 \tag{4.12}$$

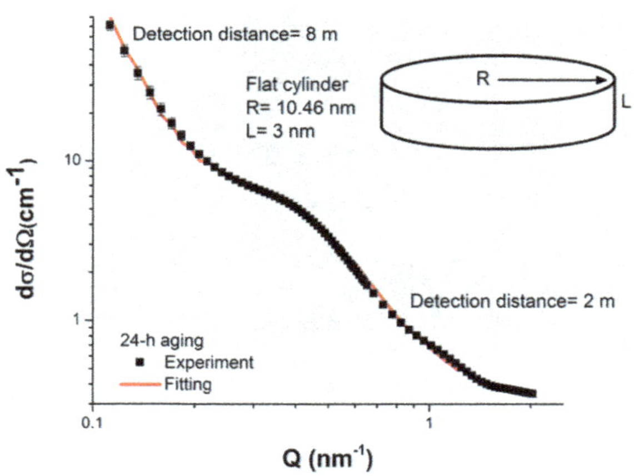

Fig. 4.9 The scattering curve plotted with the integrated one-dimensional scattering pattern

$I_{(0)}$ is the intensity when $q = 0$ and is used for the determination of the molecular weight of particles in diluted solutions [31, 32]. R_g and $I_{(0)}$ can be determined from the slope and the intercept of $I_{(0)}$ versus q^2 plot which is also known as the Guinier plot [35].

The log-normal size distribution $N(R)$ is expressed for spherical scatters as follows [33]:

$$N(R) = \frac{N_p}{R\sigma\sqrt{2\pi}} \exp\left[-\frac{(\ln(R/R_m)^2}{2\sigma^2}\right] \tag{4.13}$$

where N_p is the number density of precipitates, R_m is the mean radius and σ is the width parameter. For the assumption of a diluted spherical particle system with a homogenous $N(R)$, the mth moment of the size distribution, $\langle R^m \rangle$, is represented as:

$$\langle R^m \rangle = \int_0^\infty R^m N(R) dR / \int_0^\infty N(R) dR \tag{4.14}$$

While the first moment $\langle R \rangle$ refers as to a mean cluster radius, the third moment $\langle R^3 \rangle$ is used for the calculation of the cluster volume fraction:

$$f_p = \frac{4}{3}\pi \langle R^3 \rangle N(R) \tag{4.15}$$

4.3 Atom Probe Tomography (APT)

4.3.1 Development of APT

The first all metal atom probe field ion microscope was invented in Pennsylvania State University in 1967. It was a 1D analytical instrument, which was called atom probe field ion microscope (APFIM). The collection speed was hundreds of atoms per day and field of view (FOV) was approximately one nanometer wide. The APFIM is combined with a field ion microscope (FIM) and a time-of-flight mass spectrometer (ToFMS) with a single ion detection sensitivity. The first tomographic analysis was presented in 1973 with the ability to record hundreds of atoms per minute with a 15-nm-wide FOV. During the past decades, local electrode atom probe (LEAP) has been well developed. Besides, related techniques like the scanning atom probe, and sample preparation technique like focused ion beam (FIB) sample preparation are also promoted. Nowadays, a LEAP can collect millions of atoms per minute with up to 200-nm-wide FOV. A 3D local electrode atom probe tomography at RWTH Aachen University is shown Fig. 4.10. In LEAP, a local electrode is added close to the sample. The LEAP 4000X HR is equipped with a local electrode, a wide-angle reflection, a high-speed delay line detector system as well as an ultrafast laser

Fig. 4.10 3D local electrode atom probe (CAMECA, LEAP 4000X HR) at RWTH Aachen

with a wavelength of 355 nm. Comparing to the conventional atom probes, such an instrumental design provides fast data acquisition rates (up to 2 Mio ions/min), large analysis volumes and high compositional accuracy [36]. Large volumes containing up to several hundred million of atoms can be probed within a few hours and the impurity concentrations as low as few tens of ppm can be detected.

Figure 4.11 shows the varied applications of atom probe in materials science. APT provides three-dimensional atom-by-atom imaging of materials with a uniquely powerful combination of spatial and chemical resolution. APT has the ability to analyse composition information down to atomic scale. Other characterization techniques, like TEM, can determine physical structure of materials at sub-nanometre levels, however, only APT can identify and quantify composition of materials in three dimensional at such a scale. Moreover, APT also has high analytical sensitivity. With all the above strengths, APT can be applied in the atomic segregation in grain boundaries, the formation of rich solute clusters, composition variation in nano-grains and precipitates [37].

4.3.2 Principle of APT

The basic working principle of local electrode atom probe tomography is shown in Fig. 4.12 [38], where the electrode is positioned between the sample and the detector. The basic principle for atom probes is field evaporation, which means a high electric field is applied and the atoms at the sample surface are, therefore, ionized at cryogenic temperatures [38]. To evaporate the atoms, an electric field strengths of 10^{10} V/m or greater is required so that electrons can be stripped from atoms on the surface by a tunnelling process [38]. To achieve this, the field needs to be applied to a very sharp tip, requiring the samples to be needle-shaped with a tip radius of tens of nanometres. The position-sensitive detector records the lateral hit positions of ions and the depth position is determined regarding the accumulation sequences of the atoms. As the

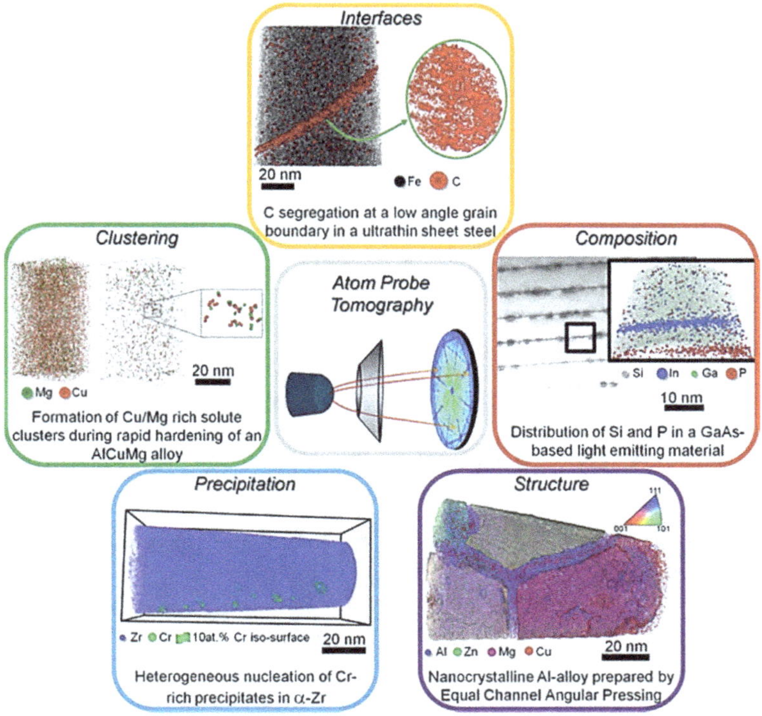

Fig. 4.11 Typical applications of atom probe tomography in materials science [37]

sample evaporating, the position of one atom in x and y is determined from its hit position on the detector and its z-position, which is the depth into the sample along the longitudinal axis, is determined from the sequence of evaporation events, allowing reconstructing a map of the constituent atoms of the sample in 3D [38]. The identity of atom is determined by time-of-flight mass spectrometry, measuring the time that takes for the ions traveling from the sample to the detector [38]. The ions are accelerated by the electric field E and the electric field at the sample apex is uniform, all ions that with the charge state n exhibit the same force. The potential energy for the ion before evaporation is $E_p = neU$, where e is electron charge, n is the ionization number, and U is the voltage [38]. It is assumed that all the potential energy is converted to kinetic energy $E_k = 1/2 \cdot mv^2$, where v is the velocity of ion and m is the mass of ion. Taking $E_p = E_k$, then

$$\frac{m}{n} = \frac{2eU}{v^2} \tag{4.16}$$

The time of flight can be written as

$$t = \frac{d}{v} \tag{4.17}$$

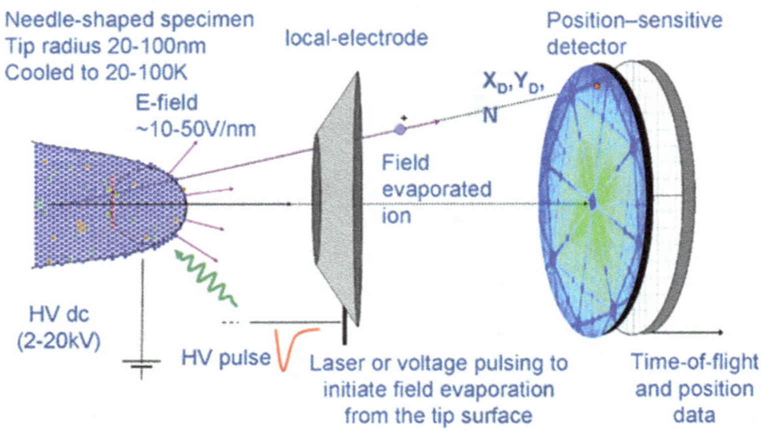

Fig. 4.12 Basic working principle of local electrode atom probe tomography [38]

where d is distance to detector. Therefore, finally the mass-to-charge ratio of evaporated ions can be obtained [38].

$$\frac{m}{n} = 2eU\left(\frac{t}{d}\right)^2 \tag{4.18}$$

4.3.3 Sample Preparation

Electropolishing using a basic loop method is shown in Fig. 4.13a [36]. In the electropolishing of APT tip preparation using the loop method in a chemically active solution with an applied voltage, the edges of the materials are more subjected to the solution and are preferentially attacked [36]. Therefore, the electro-polishing may refine the blank into a sharp needle-shaped specimen. Electropolishing is a convenient and effective way to prepare the atom probe samples and usually performed in two stages, (i) rough polishing; (ii) precise polishing. The aim of rough polishing is to eliminate the material from the sample in a quick way by using a strong electrolyte [36]. In the basic loop method, a drop of electrolyte is suspended in an ~3 mm diameter Pt loop. This loop acts as the counter electrode of the electropolishing cell. The specimen is then carefully inserted into the centre of the electrolyte-containing loop and voltage is applied permanently. Progress in the formation of a tip is monitored by occasionally pausing and viewing the specimen under a high-resolution optical microscope [36]. The curvature of the tip decreases continuously, so there is a high risk that the tip is over-polished and vanishes. In order to prevent the apex of specimen from over-polishing, a voltage-pulsed loop method is applied. The region of tip is partially immersed into the electrolyte. The power supply of the electropolishing

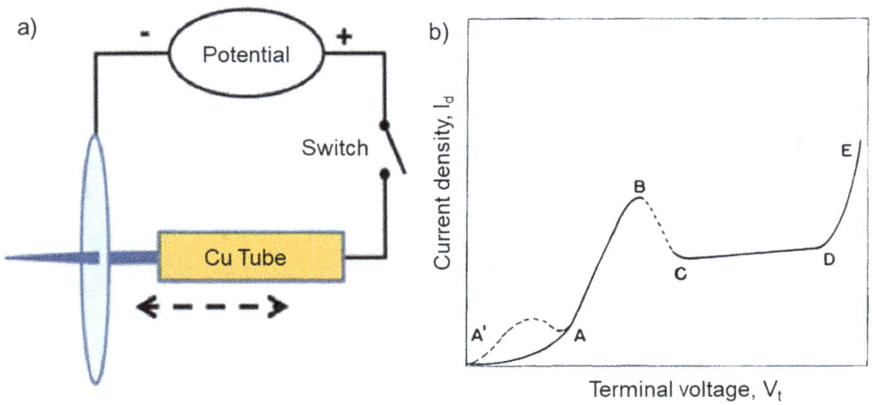

Fig. 4.13 **a** Electropolishing using a basic loop method [36]. **b** Voltage versus current density for a typical electropolishing solution [41]

setup is well designed and equipped with a pulse switch [36]. During the precise polishing, press the pulse button to switch on voltage supply intermittently. The tip part of specimen is slowly sharpened. The curvature of the tip should also be measure via high-resolution optical microscope occasionally to avoid the vanishing of sharp tip [36].

The applied voltage during the electro polishing is an important parameter (Fig. 4.13b). It reveals the relationship between current density and the terminal voltage, $i_d(V_t)$. As can be seen from the current density—voltage diagram, the curve can be divided into three main parts, namely A-B, C-D and D-E. The dashed line between A' and A is observed in certain conditions, due to the specific material and electrolyte setup, which is of less importance to the electropolishing procedure. From point A to point B, the surface of specimen experience etching and almost conserves its mechanically worked appearance. This stage can also be named as pitting stage for the signs of pitting [39]. Etching generally leads to the removal of material in a selective way, which is related to the crystallographic orientation, composition and microstructure. Under the insufficient voltage range, a solid, brown film is concomitantly formed and further destroyed at higher current density. Before the plateau (C-D) is reached, curve B-C experiences a rapid decrease, the reason behind this phenomenon is the unstable electrical characteristic, fluctuations in both voltage and current density [41]. Then the polishing effect is observed between C and D. The current density increases slightly over a wide range of voltage, forming a plateau [41]. Polishing tends to be insensitive to the microstructural difference, and removes material more rapidly from rough sites, resulting in a smoother surface. The differences between etching and polishing deserve laid more emphasis on Melmed [40], because the aim of electropolishing is to make the specimen smoothly tapered and cylindrically symmetric. The best result for polishing is obtained near point D [40]. The first bubbles of gaseous oxygen appear on the anode at point D, starting the region "gas evolution" [41]. When the voltage continues increasing to E after the

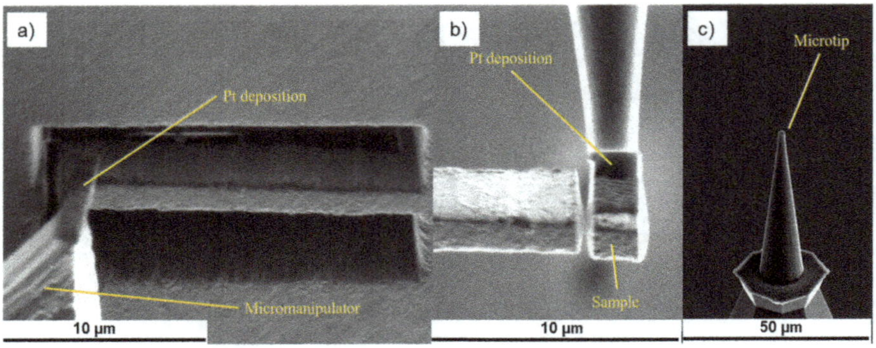

Fig. 4.14 Using FIB lift out technique to prepare an APT sample tip **a** Micromanipulator touched the undercut slice with Pt deposition; **b** Sample cut from the slice on one microtip; **c** A schematic of the microtip

point D, pitting occurs again because of contact with oxygen bubbles trapped on the surface [41]. According to the discussion above, a conclusion could be drawn that an electropolishing process is controlled by maintaining a constant voltage rather than a constant current density [41].

An alternative specimen preparation method for APT measurement is FIB milling. The samples are prepared with focused ion beam (FEI Helios Nanolab 660i Dual Beam System) in own-related work presented in the next chapters as case studies. The method used is lift-out and angular milling, as shown in Fig. 4.14. The lift-out method is a helpful and effective method for the preparation of APT specimens. The specimen slice is lifted out and manipulated to the top of a microtip array using the micromanipulator. Generally, the length of the specimen slice should be large enough to be cut into several tips. Several pieces of the specimen slice are prepared on the microtip and also subject to annular milling to sharpen the tip. The sample then is milled using Ga ions to form the needle shape of the sample tip. After finishing the first step of angular milling, the inner diameter is decreased employing decreasing current density in four steps, as shown in Fig. 4.15.

The step-by-step combinations of inner diameter and current density applied in the investigated materials are: (i) 0.79 nA with 0.75 μm; (ii) 0.43 nA with 0.5 μm; (iii) 0.23 nA with 0.3 μm and (iv) 80 pA with 0.1 μm. The typical tip radius and the shank angle of the materials tip is shown in Figs. 4.15e and 4.16.

4.3.4 3D Atoms Map Reconstruction

Reconstructing the original positions of the ions is very important to the generation of the 3D map of the sample. To observe the location of clusters, segregation and precipitation, accurate position reconstruction is required and to achieve precise chemical composition, the mass spectrum should be accurate. The identity of individual ions

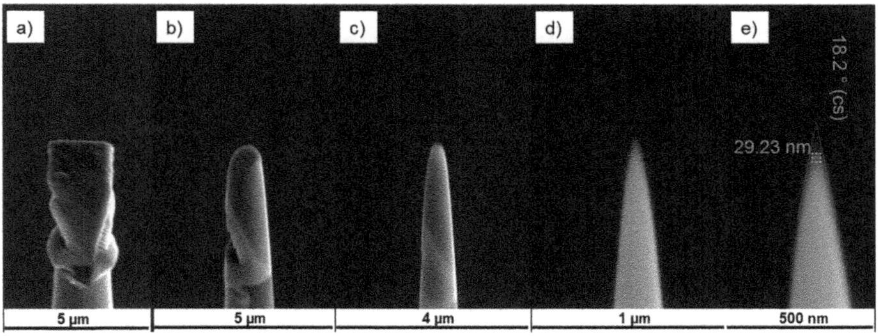

Fig. 4.15 **a** The initial sample subject to angular milling; the sample images recorded after **b** 1.5 μm milling (0.79 nA), **c** 0.75 μm milling (0.79 nA), **d** 0.3 μm milling (0.23 nA), **e** 0.1 μm (80 pA) and Ga-decontamination

Fig. 4.16 A prepared sample on the top of the silicon coupon

is determined by measuring their flight times to reach the detector. The precise concentration profile or proxigram can be obtained only when the reconstructions are constructed correctly [38]. A mass calibration must be carefully completed with appropriate range selections [38]. Ranging the elemental profiles not only provides the identification of a peak, but also impacts the compositional accuracy and determines the analysis volume of the nanostructures in the materials. A typical mass spectrum after finishing all the steps is shown in Fig. 4.17.

An accurate reconstruction can construct the 3D map of the sample with correct shape and tip radius. The shank angle can be measured by the FIB system after the preparation of the sample. The tip radius can also be adjusted by analysing the voltage profile using the software. This can provide a better and more precise reconstruction.

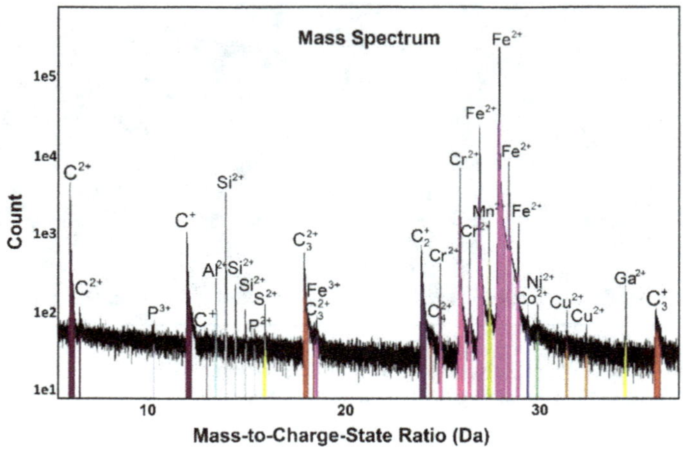

Fig. 4.17 A mass spectrum obtained from APT data analysis

Quenching and partitioning (Q&P) steels are one of the most promising candidates in the 3rd generation AHSS. Carbon partitioning between martensite and austenite plays an important role to stabilize austenite in Q&P steels [42]. Figure 4.18 shows an example of elemental partitioning features in a Q&P steel Fe-1.36C-3.3Si-2.6Mn (at.%).

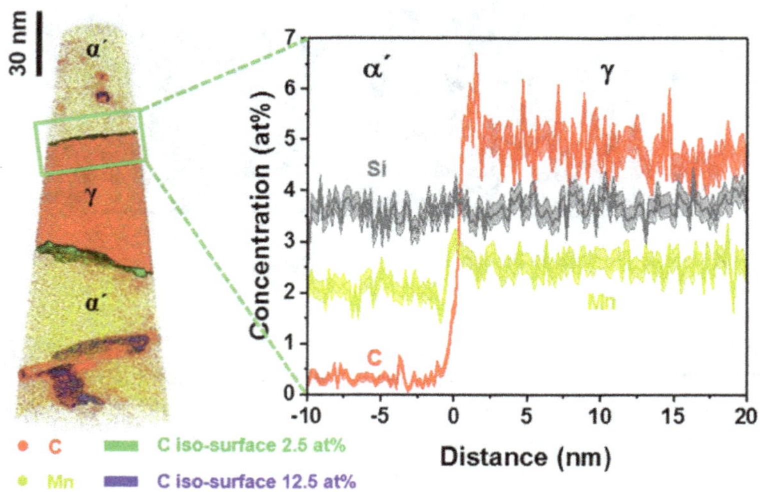

Fig. 4.18 APT 3D reconstruction map [42]

4.4 High-Resolution Transmission Electron Microscopy (HRTEM)

4.4.1 Introduction to Electron Crystallography

In the year 1938, Pinsker and Vainshein started to build up cameras to capture the electron diffraction patterns of crystal materials in Moscow. It was a precocious idea because neither theories nor techniques of electron diffraction analysis had been developed by that time. Ten years later, the first Fourier map of a crystal structure based on electron-diffraction data of $BaCl_2 \cdot H_2O$ was achieved, and the technique was developed into a comprehensive and self-sustaining research field of materials characterization. That is the starting point of electron diffraction (ED) analysis. The first high-resolution TEM, Siemens Elmiskop, was assembled in 1954. In year 1956, Hirsch et al. successfully developed foil sample preparation method of a number of different steels, including steels [43]. Precipitation in Q&T low alloy steels is one of the first areas studied using this new equipment. The theories of kinetics and dynamic diffraction contrast were developed by a period of years, which allows the details of structure features in the materials to be observed in foil TEM specimens, such as grain boundaries, dislocations and precipitates to be interpreted.

Over the past decade, high-resolution TEM has been widely correlated with other advanced materials characterization methods to interpret the crystal structure in accordance with other necessary information [43]. The correlative use of APT and HRTEM provides joint chemical and structural insights to the fundamental scientific questions in materials science [44]. Electron crystallography can also be used to extend the range of samples amenable to structure analysis beyond those that can be studied by single crystal X-ray diffraction. Electrons can be focused by magnetic lenses to form an image, HRTEM images of crystals can be obtained, while X-ray imaging is not possible. However, although electron crystallography has many unique advantages, it has not been widely used as a standard technique for crystal structure determination of inorganic crystals. HRTEM images are usually not directly interpretable without image processing because of optical distortions, crystal tilt and multiple scattering.

4.4.2 Principles of TEM/STEM

TEM is one of the principal analytical devices employed in the research and development of metallic materials. Scanning transmission electron microscope (STEM) is a type of TEM that features a scanning function. STEM is superior in its capacity for microstructural observation and elemental analysis of local regions of a specimen. With STEM, it is possible not only to directly observe the atomic structure of the grain boundary, phase interface, and precipitate, but also to observe an image of the specimen where the contrast is directly related to the atomic number. When synchronized with the electron beam scanning function, elemental mapping can be performed. Aberration-corrected STEM has a built-in mechanism for correcting the undesired spherical aberration of the irradiation lens system to condense the beam of electrons. Both image resolution and spatial resolution in elemental analysis improve to 0.1 nm or less. Therefore, it is expected that the aberration-corrected STEM will allow for sub-nanometre-level analysis of nano-sized precipitates.

In precipitation strengthening for steel materials, the precipitate size that provides the better strengthening-ductility ratio is normally in the range of several to less than ten nanometres [45]. Therefore, in order to secure an adequate amount of precipitation strengthening by means of precipitation control, information about the distribution, shape, structure, and composition of nano-sized precipitates is indispensable. For example, Fig. 4.19 shows the microstructure of a nano-precipitate in a dual phase (DP) steel from relatively low to very high magnification corresponding to atomic-level resolution [46]. This means that aberration-corrected STEM makes it possible to analyse the condition of dispersion and the composition of nano-sized precipitates in steel. Recent technological improvements allow to effectively investigate the size, morphology and chemistry of such nano-scale precipitates, by using scanning transmission electron microscopy in combination with electron energy loss spectroscopy (EELS) [47]. Due to the broad range of inelastic interactions of the high energy electrons with the specimen atoms, ranging from phonon interactions to ionisation processes, electron energy-loss spectroscopy offers unique possibilities for advanced materials analysis [46]. It can be used to map the elemental composition of a specimen, but also for studying the physical and chemical properties of a wide range of materials [46].

4.4.3 Electron Microscopy and Electron Diffraction

Electron beams are sensitive to the variations of electrostatic potential in crystals. When electrons penetrate a crystal, the information about the potential distribution of the crystal can be gathered. The electrostatics potential is a three-dimensional function that is rather smooth between the atoms but rises sharply to high values where the atomic nuclei are. Thus, determining the potential distribution in a crystal to high-resolution is equivalent to finding the atomic positions. Electron diffraction in HRTEM is also suitable for orientation microscopy, based on two principally different modes, Kikuchi patterns and spot patterns. Transmission Kikuchi patterns (Fig. 4.20) show many similarities with electron backscattering diffraction (EBSD) patterns, in particular concerning the necessity of an incoherent scattering step to create an electron source in the crystal. Spot patterns (Fig. 4.21) are formed exclusively by coherent scattering which requires less interaction with the material. Therefore, the beam dose can be lowered, and the spatial resolution may be better [48].

Transmission Kikuchi patterns (TKP) are created by the same physical principles as backscatter Kikuchi patterns, i.e. incoherent, quasi-elastic scattering of the primary beam electrons followed by coherent scattering at the lattice [48]. A typical diffraction

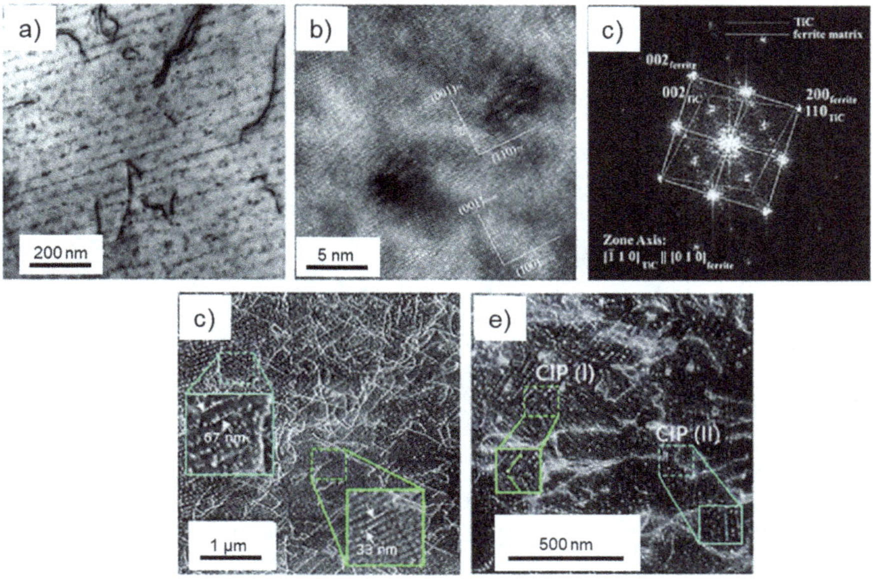

Fig. 4.19 Microstructure of the precipitation hardened dual phase steel: **a** TEM bright field image showing interphase precipitation of nano-sized titanium carbides; **b** HRTEM image of a nanometre-sized carbide taken from the area in (**a**); **c** The corresponding FFT diffractogram obtained from (**b**); **d** STEM image of CuTi microalloyed DP steel after tempering at 500 °C for 3 h; **e** STEM image of precipitation morphology of interphase precipitation carbides [46]

pattern and a schematic drawing of its geometry are displayed in Fig. 4.20. Kikuchi patterns have two important advantages with respect to other TEM-based diffraction techniques for orientation microscopy. The first is the high orientation resolution, i.e. the high accuracy with which an orientation can be calculated from the diffraction pattern [48]. The high angular sensitivity of the Kikuchi technique is at the same time also the reason for the most serious disadvantage of the technique, its sensitivity to lattice defects. A second advantage of Kikuchi patterns is the rareness of 180° orientation ambiguities [48]. Rather, there are two crystal orientations possible that are related by a 180° rotation about the pattern zone axis. In Kikuchi patterns, due to the relatively large solid angle of scattering (up to 20°) there are always diffraction vectors from several zone axes available which makes orientation determination unequivocal [48].

The classical and probably still most widely used form of diffraction pattern in TEM is the selected area electron diffraction (SAED) pattern, as is shown in Fig. 4.21 [48]. The SAED pattern is produced by illumination of the thin foil sample with a parallel electron beam (i.e. a plane wave front). The diffracted beams are focused into sharp spots in the back focal plane of the objective lens [48]. In order to limit the area of the sample that contributes to the diffraction pattern a selector aperture is introduced into the intermediate image plane. The SAED pattern has to be understood

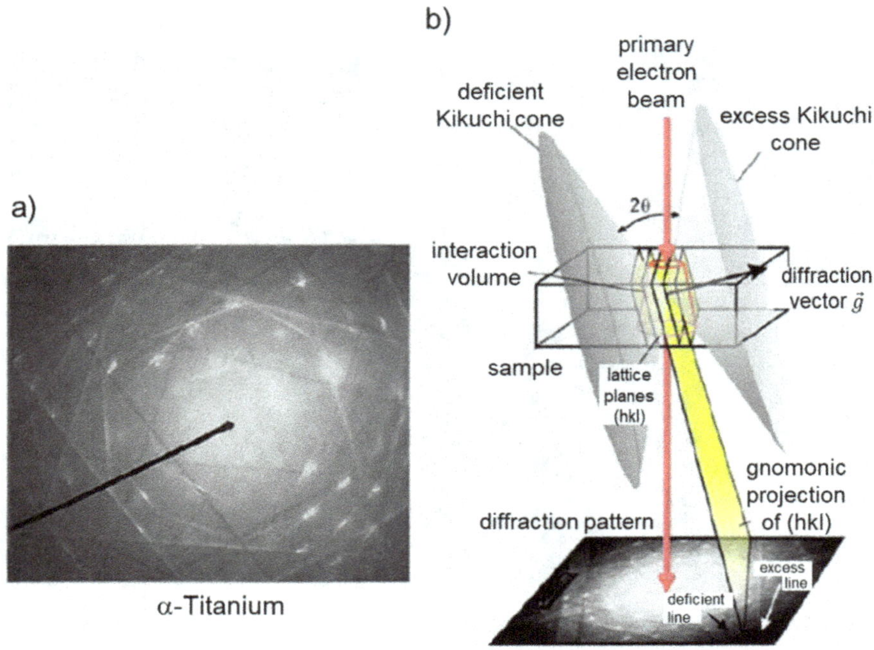

Fig. 4.20 **a** Typical transmission Kikuchi pattern obtained at 200 keV from α-titanium; and **b** Schematic drawing of the geometry of the TKP origination [48]

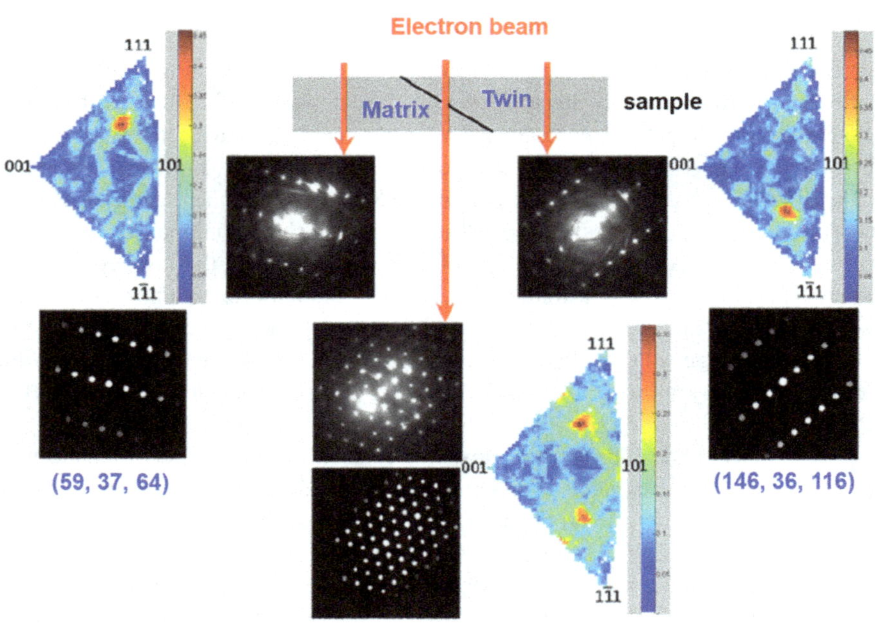

Fig. 4.21 Template matching applied to solve the overlapping patterns of twin and matrix in a copper sample [48]

as the central projection of a spherical cut through the reciprocal lattice into real space [48].

4.4.4 Sample Preparation and Measurements of HRTEM

The STEM samples can be prepared by electro polishing and partially by FIB cutting. The electropolishing includes three steps: (i) cut the material discs with a diameter of 3 mm; (ii) mechanically grind the discs to a thickness of ~100 μm. Thin disc samples with different strain levels are cut from interrupted tensile test samples with the surface perpendicular to the tensile axis, i.e. the observation direction (normal to the sample surface) is always along the tensile axis. (iii) Electropolishing is carried out in a Struers Tenupol twin-jet device at −30 °C and 10 V using a solution of 30% nitric acid in methanol. In FIB system, Ga ions are applied to remove the material and cut a lamella. Then, the lamella is transferred and fixed to a grid or half grid, with the "lift-off" technique. The advantages of using a FIB for TEM specimen preparation are manifold [49]: (i) there is no other technique with which such a precise selection of the target area can be reached, i.e. lamellae can be placed with an accuracy of ~20 nm. (ii) FIB preparation is fast and reliable, i.e. in as little as 20 min, and within a maximum of 2–4 h, specimens can be prepared of an almost unlimited range of materials. The high-resolution (HR) imaging was performed with an aid of FEI Titan G2 80–200 Cs probe-corrected STEM [50], equipped with a high-brightness Schottky field emission electron gun operated at 200 kV (HRSTEM) and FEI Titan 80–300 field emission gun transmission electron microscope equipped with an imaging spherical aberration corrector operated at 300 kV [51].

For answering many materials questions, correlative chemical and crystallographic knowledge at the nano-scale is required. The complementary nature of TEM and APT suggests that both techniques should be subsequently applied to the same sample in order to combine their strengths. A versatile experimental setup based on a modified, commercial, high-angle single-tilt TEM retainer was developed by [49], which enables correlative FIB, scanning electron microscopy (SEM), TEM and APT, can be used to prepare the sample used for the TEM-APT correlative observation. The core part of the experimental setup consists of a modified replica of a JEOLEM-21311HTR high-angle single-tilt TEM retainer for a JEOL JEM-2200FS microscope. The front/end half of the grid-holding section is removed and the front ends are tapered (Fig. 4.22a) [49].

The electro-polished grid is placed into the grid holder using a vacuum tweezer (Fig. 4.22f). For FIB milling the grid holder is inserted into an adapter which is aligned vertically into the universal mounting base (Fig. 4.22c) in a FEI Helios Nanolab 600i dual-beam FIB instrument [49]. By a standard lift-out procedure, four samples are deposited on electropolished grid posts. During annular milling of the samples, the stage is tilted to 52° so that the grid holder's tilt axis is parallel to the ion beam [49]. For single-tilt TEM measurements the grid holder is directly inserted into a JEOLJEM-2200FS microscope operating at 200 kV. The clamping

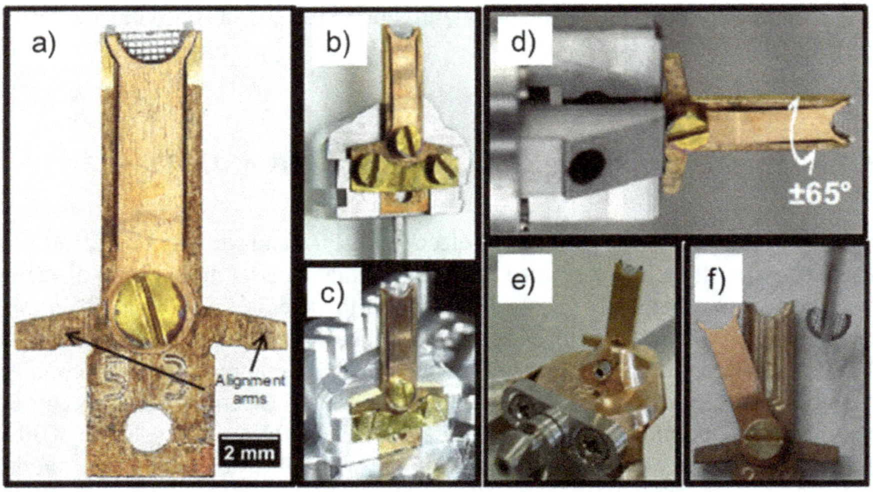

Fig. 4.22 The "grid holder", core part of the experimental setup: **a** the grid holding section; **b** the adapter employed during FIB milling; **c** Experimental setup during FIB sample preparation; **d** The grid holder fits directly into a JEOL TEM single-tilt holder; **e** Modification of a Cameca puck specimen assembly that allows the insertion of the grid holder into the LEAP instrument; **f** Insertion/extraction of the grid using a vacuum "tweezer" holding tool [49]

mechanism of the grid holder makes the transfer of the samples to any other standard TEM retainer using a vacuum holding tool, a straight-forward procedure (Fig. 4.22f) [49]. This enables measurement of APT tips in virtually all TEMs. For double-tilt measurements described here, the grid is transferred into a double-tilt TEM retainer and then back into the grid holder to conduct the APT measurements [49].

4.5 Summary

In order to fully understand the phase transformation and strengthening mechanisms in nanostructured metallic materials, a combined multi-scale characterization approach is necessary (Fig. 4.23). High-resolution TEM enables the study of the microstructural characteristics at atomic scales, such as atoms ordering, defects, and interfaces with detailed crystallographic information. APT offers the nearly atom-by-atom 3D reconstruction of the material structure for a volume about 10^6 nm^3. It provides the great opportunity for the precise chemical composition analysis in specimens in a local chemical atmosphere, particularly the elemental partitioning features at specific locations, such as grain boundaries, etc. SYXRD and SANS are very helpful and powerful tools to gain an overview of the nanostructure in a larger volume. It supports to bridge the characterization and understanding of the materials from atomic scale, up to nano-, micro- and further to macro-scale. In order

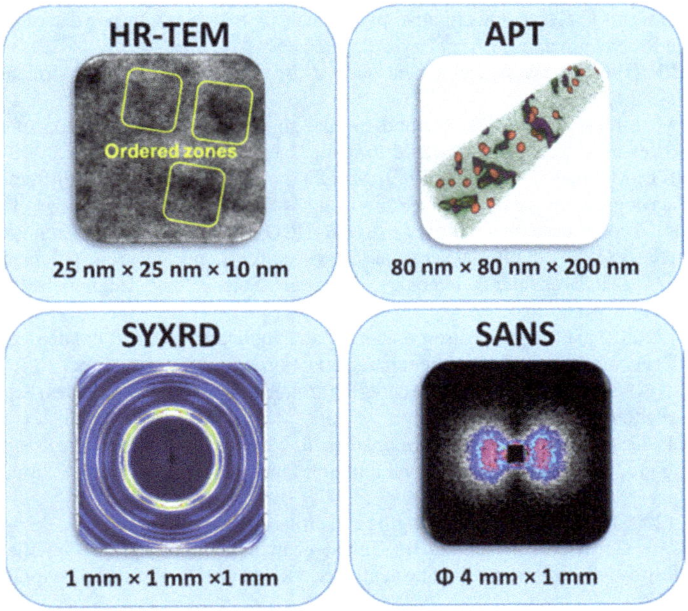

Fig. 4.23 Combined nanostructure characterization techniques and their typical detection volume

to study the structure-property relationship in metallic materials, the scale-bridging experiments are highly important.

References

1. National Synchrotron Radiation Research Centre (2018) Introduction of synchrotron source. https://www.nsrrc.org.tw/chinese/lightsource.aspx. (Online)
2. Engler O, Randle V (2009) Introduction to texture analysis: macrotexture, microtexture, and orientation mapping, 2nd edn. CRC Press, Baton Rouge
3. Willmott P (2011) An introduction to synchrotron radiation: techniques and applications, 1st edn. John Wiley & Sons, Hoboken
4. Mobilio S, Boscherini F, Meneghini C (2015) Synchrotron radiation: basics, methods and applications, 1st edn. Springer, New York
5. Hwu Y, Margaritondo G (2021) Synchrotron radiation and X-ray free-electron lasers (X-FELs) explained to all users, active and potential. J Synchrotron Radiat 28:1014–1029
6. Gottstein G (2004) Physical foundations of materials science, 1st edn. Springer, New York
7. Fultz B, Howe J (2013) Diffraction and the X-ray powder diffractometer. In: Transmission electron microscopy and diffractometry of materials, 3rd edn. Springer, New York
8. Ma Y, Sun B, Schökel A, Song W, Ponge D, Raabe D, Bleck W (2020) Phase boundary segregation-induced strengthening and discontinuous yielding in ultrafine-grained duplex medium-Mn steels. Acta Mater 200:389–403
9. Dippel A, Liermann H, Delitz J, Walter P, Schulte-Schrepping H, Seeck O, Franz H (2015) Beamline P02.1 at PETRA III for high-resolution and high energy powder diffraction. J Synchrotron Radiat 22:675–687

10. Hill R, Howard C (1987) Quantitative phase analysis from neutron powder diffraction data using the Rietveld method. J Appl Crystallogr 20:467–474
11. Rietveld H (1969) A profile refinement method for nuclear and magnetic structures. J Appl Crystallogr 2:65–71
12. Ferrari M, Lutterotti L (1994) Method for the simultaneous determination of anisotropic residual stresses and texture by X-ray diffraction. J Appl Phys 76:7246–7255
13. Lutterotti L, Matthies S, Wenk H (1999) MAUD (material analysis using diffraction): a user friendly Java program for Rietveld texture analysis and more. In: Proceedings of the twelfth international conference on textures of materials (ICOTOM-12). NRC Research Press, Ottowa
14. Hammouda B (1995) A tutorial on small-angle neutron scattering from polymers. National Institute of Standards and Technology, Materials Science and Engineering Laboratory, Gaithersburg
15. Grillo I (2008) Small-angle neutron scattering and applications in soft condensed matter. In: Borsali R, Pecora R (eds) Soft matter characterization. Springer, New York
16. Loveley S (1984) Theory of neutron scattering from condensed matter: nuclear scattering, 1st edn. Clarendon Press, Oxford
17. Perrard F, Deschamps A, Bley F, Donnadieu P, Maugis P (2006) A small-angle neutron scattering study of fine-scale NbC precipitation kinetics in the α-Fe-Nb-C system. J Appl Crystallogr 39:473–482
18. Sears V (2006) Neutron scattering lengths and cross sections. Neutron News 3:26–37
19. Chang C, de Geuser F, Banhart J (2015) In-situ characterization of β'' precipitation in an Al-Mg-Si alloy by anisotropic small-angle neutron scattering on a single crystal. J Appl Crystallogr 48:455–463
20. Squires G (2010) Introduction to the theory of thermal neutron scattering, 3rd edn. New York, Cambridge University Press, New York
21. Shah A, Shutt R, Smith K, Hack J, Neville T (2021) Neutron studies of Na-ion battery materials. J Phys Mater 4:042008
22. Song W, Bogdanovski D, Yildiz A, Houston J, Dronskowski R, Bleck W (2018) On the Mn-C short-range ordering in a high strength high-ductility steel: small angle neutron scattering and ab initio investigation. Metals 8:44
23. Kohlbrecher J (2014) SASfit: a program for fitting simple structural models to small angle scattering data. https://kur.web.psi.ch/sans1/sasfit/sasfit.pdf. (Online)
24. Zhu Y (2005) Small-angle neutron scattering. In: Zhu Y (ed) Modern techniques for characterizing magnetic materials. Springer, Boston
25. King S (1999) Small angle neutron scattering. In: Pethrick RA, Dawkins JV (eds) Modern techniques for polymer characterization. Wiley & Sons, New York
26. Miller M, Wirth B, Odette G (2003) Precipitation in neutron-irradiated Fe-Cu and Fe-Cu-Mn model alloys: a comparison of APT and SANS data. Mater Sci Eng A 353:133–139
27. Karaman I, Sehitoglu H, Maier H, Chumlyakov Y (2001) Competing mechanisms and modeling of deformation in austenitic stainless steel single crystals with and without nitrogen. Acta Mater 49:3919–3933
28. Turk C, Leitner H, Kellezi G, Clemens H, Gan W, Staron P, Primig S (2016) Impact of the B2 ordering behavior on the mechanical properties of a FeCoMo alloy. Mater Sci Eng A 662:511–518
29. Shankar P, Sundararaman D, Ranganathan S (1998) Clustering and ordering of nitrogen in nuclear grade 316LN austenitic stainless steel. J Nucl Mater 254:1–8
30. Radulescu A, Szekely N, Appavou M (2015) KWS-2: small angle scattering diffractometer. JLSRF 1:29
31. King S (1995) Small angle neutron scattering. ISIS Facility, STFC Rutherford Appleton Laboratory
32. Pedersen J (1994) Determination of size distribution from small-angle scattering data for systems with effective hard-sphere interactions. J Appl Crystallogr 27:595–608
33. Keiderling U, Wiedenmann A, Srdic V, Winterer M, Hahn H (2000) Nano-sized ceramics of coated alumina and zirconia analyzed with SANS. J Appl Crystallogr 33:483–487

34. Staron P, Jamnig B, Leitner H, Ebner R, Clemens H (2003) Small-angle neutron scattering analysis of the precipitation behaviour in a maraging steel. J Appl Crystallogr 36:415–419
35. Reimers W, Pyzalla A, Schreyer A, Clemens H (2008) Neutrons and synchrotron radiation in engineering materials science: from fundamentals to material and component characterization, 1st edn. Wiley-VCH, Weinheim
36. Larson D, Prosa T, Ulfig R, Geiser B, Kelly T (2013) Local electrode atom probe tomography: a user's guide. Springer, New York
37. Gault B, Moody M, Cairney J, Ringer S (2012) Atom probe microscopy, 1st edn. Springer, New York
38. Gault B, Moody M, de Geuser F, Tsafnat G, Fontaine A, Stephenson L, Haley D, Ringer S (2009) Advances in the calibration of atom probe tomographic reconstruction. J Appl Phys 105:034913
39. Landolt D (1987) Fundamental aspects of electropolishing. Electrochim Acta 32:1–11
40. Melmed A (1991) The art and science and other aspects of making sharp tips. J Vac Sci Technol B 9:601–608
41. Metz F (1960) Electropolishing of metals. Doctoral thesis, Lowa State University
42. Bleck W, Brühl F, Ma Y, Hasse C (2019) Materials and processes for the third-generation advanced high strength steels. METEC & 4th ESTAD, Düsseldorf
43. Hirsch P, Horne R, Whelan M (1956) Direct observations of the arrangement and motion of dislocations in aluminium. Philos Mag 1:677–684
44. Herbig M, Choi P, Raabe D (2013) Combining structural and chemical information on the nanometer scale by correlative TEM and APT. Microsc Microanal 19:948–949
45. Suwanpinij P, Dickert H (2014) Application of vanadium precipitation for lower rolling force and enhanced strength of hot strip steel. Arch Mater Subj 70:39–44
46. Li C, Chen C, Tsai S, Yang J (2019) Microstructure characterization and strengthening behavior of dual precipitation particles in CuTi microalloyed dual-phase steels. Mater Des 166:107613
47. Gázquez J, Sanchez-Santolino G, Biskup N, Roldan M, Cabero M, Pennycook, S, Varela M (2017) Applications of STEM-EELS to complex oxides. Mater Sci Semicond Process 65:49–63
48. Zaefferer S (2011) A critical review of orientation microscopy in SEM and TEM. Cryst Res Technol 46:607–628
49. Herbig P, Choi D (2015) Raabe, combining structural and chemical information at the nanometer scale by correlative transmission electron microscopy and atom probe tomography. Ultramicroscopy 153:32–39
50. Kovács A, Schierholz R, Tillmann K (2016) FEI Titan G2 80–200 CREWLEY. JLSRF 2:43
51. Thust A, Barthel J, Tillmann K (2016) FEI Titan 80–300 TEM. JLSRF 2:41

Chapter 5
Precipitation Engineering

This chapter describes the precipitation engineering approach and its application in the high strength steels. The strengthening mechanism of precipitation hardening are explicitly discussed in Sect. 5.2.1, and the influences of precipitate characteristics, *i.e.* structure, size, number density, distribution, thermal stability, on the strengthening effects will be reviewed in Sect. 5.2.2. Own-related research work,[1,2] of precipitation engineering is then reported as application examples in Sect. 5.4.

5.1 Definition and History

Precipitation hardening is a heat treatment technique used to increase the strength of materials, including most structural alloys of aluminium, magnesium, nickel, titanium, steels and stainless steels [1]. With the theoretical and empirical base of precipitation hardening approach, precipitation engineering aims at making use of the strengthening mechanisms to design the characteristic of precipitates according to different saturating property between the matrix phase and the precipitates under different heat treatment temperatures. Precipitation usually occurs at the defect sites, *i.e.* dislocations, grain boundaries, phase boundaries, resulting in different interactions with dislocations, leading to different strengthening effects. For nano-precipitation engineering, it is desired to obtain dispersedly-distributed nano-sized precipitates for strengthening the materials. Strengthening metallic materials with precipitation dates back to 1901 [1]. It started with the research of Dr. Alfred Wilm on aluminium, when it was accidently discovered an increasement of hardness on

Own work-related examples link to the publications:

[1] Song et al. [53].

[2] Pan et al. [58].

W. Song, *Nano-Engineering of High Strength Steels*,
Topics in Mining, Metallurgy and Materials Engineering,
https://doi.org/10.1007/978-3-031-42967-5_5

the aluminium alloy after two days at room temperature. This phenomenon was termed as "age hardening". In 1938, Guiner and Preston found the clustering of atoms with X-ray scattering, which now is called Guinier–Preston zones. In year 1958, the first commercial niobium-treated steels entered the market. The very small amount of Nb addition (0.005 wt.% to 0.03 wt.%) leads to large strengthening effect combined with good toughness. In recent years, with the development of the sophisticated characterization techniques, *i.e.* high-resolution transmission electron microscopy (HRTEM), synchrotron X-ray diffraction (SYXRD), atom probe tomography (APT), small angle neutron scattering (SANS), electron channelling contrast imaging (ECCI), more detailed insights of the nano-precipitation mechanisms are revealed with the possibility to gain detailed quantitative knowledge of the materials.

5.2 Precipitation Strengthening Mechanisms

5.2.1 Precipitation Strengthening Mechanisms

Precipitation strengthening has been used extensively for a long time period, and in a wide range of metallic systems. Principally, precipitation strengthening comes to effect by producing a particulate dispersion of motion obstacles for dislocation motion. The theory of interaction process between dislocation and incorporated particles is well established in physical metallurgy. Figure 5.1 schematically shows the forces exerted on a mobile dislocation in a stressed state, which is impeded by a secondary phase particle. In this case, the balanced forces between line tension of dislocation and the resistant force resulted from particle can be expressed as [2]:

$$F = 2T sin\theta \tag{5.1}$$

where T denotes the line tension of the dislocation, F is the resistant force of particle and θ is the angle of deflection. Dislocation motion is the prerequisite of plastic deformation. Therefore, the force equation is imbalance when the deformation proceeds [2].

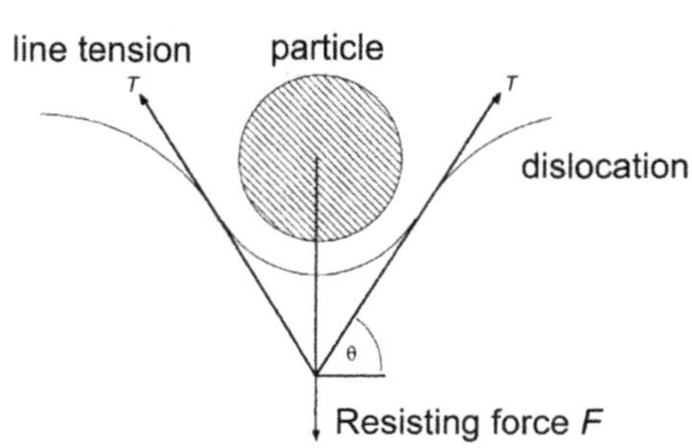

Fig. 5.1 Balance of forces acting during particle resistance to dislocation movement [2]

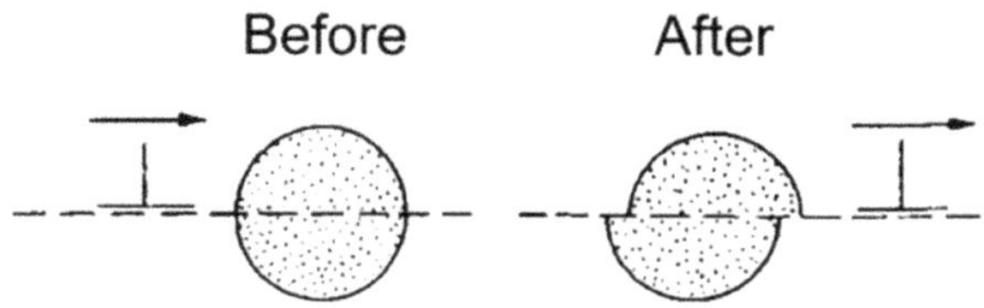

Fig. 5.2 Dislocation motion cutting through the second phase particles [2]

The operative interaction mechanism between particles and dislocations depends predominantly on the nature of the particles, *i.e.* coherency or particle deformability. In the small-particle regime the precipitates, or particles, are coherent and deformable as the dislocations cut through them as shown in Fig. 5.2 [2]. The strength increase is proportional to $r^{1/2}$, where r is the particle radius.

When the particle/matrix interface is incoherent, instead of cutting the particles, the dislocations pass by the particle, leaving a dislocation loop behind, which is called Orowan looping mechanism. The Orowan mechanism is schematically illustrated in Fig. 5.3. Under plastic deformation, either mobile dislocations cut penetrable particles and leave behind a sheared particle or in the case of impenetrable particles, plastic deformation leads to bowing of dislocations around particles, leaving behind the characteristical loops [2]. In the approaching stage, the dislocation moves close to the second phase particle by the applied force (Fig. 3a). As soon as the dislocation is pinned by the particles, extra force is required to trigger the motion of dislocation, which responds to work hardening procedure (Fig. 3b). When the force acting on the dislocation segment is large enough to bend the dislocation into a semi-circle with a critical radius r_c (Fig. 3c), the further motion of dislocation proceeds simultaneously (Fig. 3d). Additionally, for apparent coarse particles, bowing of dislocations always occurs regardless of the interface coherency or particle formability [2].

Orowan looping appears preferably in the samples in long aging conditions and it may be influenced by the change of the size, number density and coherency of the nano-precipitates to the matrix [3, 4]. A critical radius r_c is commonly used to calculate the transition from shear to loop mechanism. Assuming the critical resolved shear stress (CRSS) is equal for shearing and looping, respectively, r_c was calculated according to the following expression [5]:

$$r_c = \left(\frac{\alpha}{0.7}\right)^{2/3} \times \frac{\mu b^2}{APB}\left(\frac{f}{(1-f)^2}\right)^{1/3} \tag{5.2}$$

where μ, b and APB, represent the shear modulus, the Burger's vector and the antiphase boundary energy respectively. The strength increases by Orowan looping mechanism is proportional to r^{-1}, where r is the particle radius. Therefore, the predominant strengthening mechanism can be described as the competition, where the lower strength triggers the deformation early. The competition relation between cutting through and Orowan is schematically illustrated in Fig. 5.4 [6]. For coherent and semi-coherent particles, the strengthening mechanism transfers from cutting-off to Orowan passingby [6]. The critical stress for dislocation to cut the particles

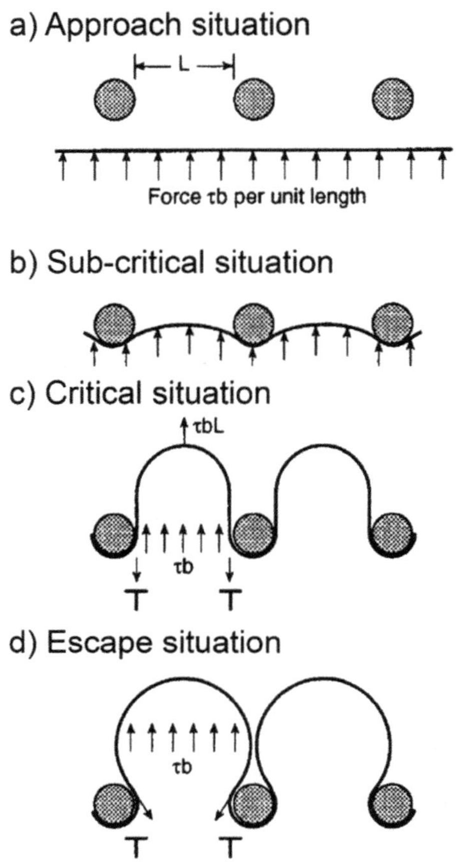

Fig. 5.3 Dislocation meets hard undeformable second phase particles: dislocation release at higher stresses may occur by Orowan looping or by cross-slip [2]

increases with the particle size, which is lower than that to pass by the particles [6]. With the elevated particle size, the critical stress to forming the dislocation loop becomes lower than that of the cutting-off mechanism. Therefore, the Orowan dislocation loop mechanism comes into effect. For non-coherent particles, the misorientation between particle and matrix is large and the lattice structure differs [6]. Dislocations has no chance to cut through the particles. The Orowan mechanism predominates the strengthening progress regardless of the particle size [6].

An example is shown in powder metallurgy Inconel 100 alloy with the chemical composition (in wt.%): 4.85 Al, 4.24 Ti, 18.23 Co, 12.13 Cr, 3.22 Mo, 0.71V,0.071 Zr, 0.02 B, 0.072C and balance Ni [5]. As shown in Fig. 5.5, Orowan loops are primarily observed in the samples with larger articles, while for smaller particles mostly shearing occurred. It is also reported that both shear and loop mechanisms may occur simultaneously. Different slip systems can be seen in Fig. 5.6a with minor Orowan loops and Fig. 5.6b reveals the shearing of the second and third phase SII and SIII, respectively [5, 7].

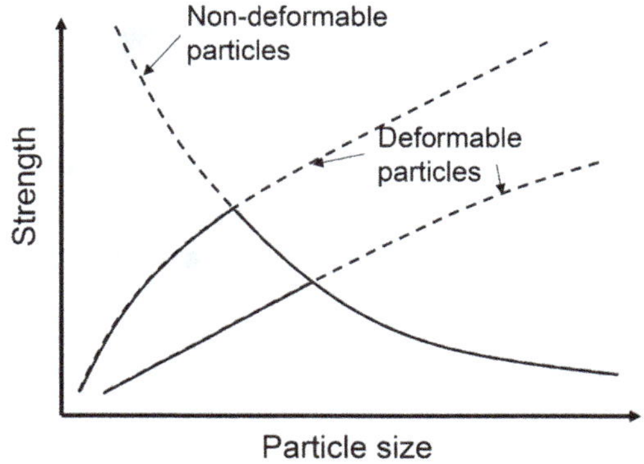

Fig. 5.4 Variation of strength with particle size, defining the deformable and non-deformable particle regime [6]

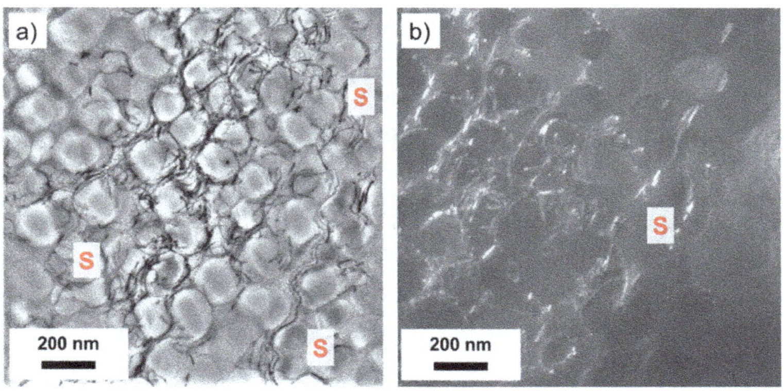

Fig. 5.5 TEM images of larger sized particles in Inconel IN100 with major Orowan loops and minor shearing, indicated as "S" **a** in bright field; **b** in dark field [5]

With the development in characterization technique, the near atomic level analysis can be conducted by APT, delivering the highest spatial resolution in 3D space. The cutting through mechanism was recently experimentally observed by Yao [8] in a high-Mn Fe-30Mn-8Al-1.3C (wt.%) steel. To induce the precipitation of kappa phases and study their evolution and phase stabilities upon ageing, the solid solution-treated sample was isothermally annealed at 600 °C for 24 h and oil quenched. The deformed microstructure with 3D chemical mapping was reconstructed, accomplished by correlative transmission electron microscopy (TEM)-APT on a 0.05-strained specimen, as presented in Fig. 5.7 [8]. In the TEM micrograph (Fig. 5.7a), two originally cuboidal or plate-like kappa phase precipitates are clearly split into two parts—one major part and another smaller one at the corner, as highlighted by blue arrows. The slight mismatch between these two parts, particularly for the right plate-like kappa phase, indicates that these two precipitates are sheared by dislocation

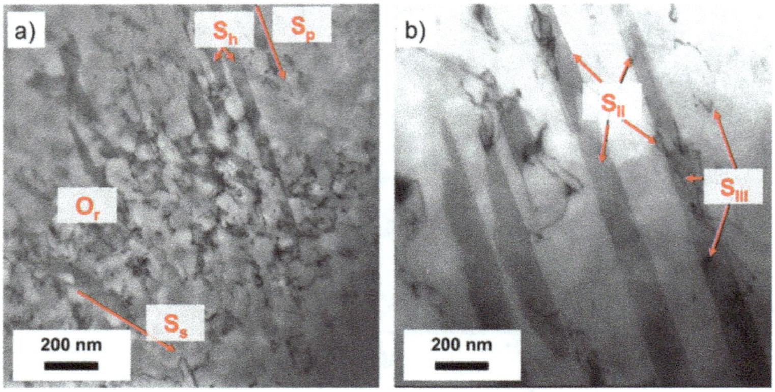

Fig. 5.6 Post mortem observations of the water quenched IN 100 material showing the shearing mechanism: **a** Two different systems are activated (principal S_P) and secondary system (S_S), super-shockley dislocations (S_h) and isolated Orowan by-passing (O_r); **b** Zoom-in image showing the secondary shearing (S_{II}) and the tertiary shearing [5]

slip [8]. The slip directions of the two precipitates are parallel to each other. Since the needle-shaped specimen is prepared orthogonally along the < 001 > direction, the slip direction can be characterized as ~45° from [001] and [010], *i.e.* close to [011]. Besides, as highlighted by yellow arrows, there are also some small fragments of the ordered precipitates, which are probably heavily fragmented by deformation. The same-site APT result imposed on top of the dark field (DF)-TEM micrograph is shown in Fig. 5.7b [8]. In addition to large kappa phases both observed in TEM and APT, there are also small, irregularly shaped fragments with no obvious visual match between TEM and APT measurements: the region of which is marked by a yellow dash circle in Fig. 5.7b). To better reveal the 3D solute distribution after deformation, further APT elemental maps and iso-concentration surfaces of different elements at varying threshold values are created (Fig. 5.7c–e) [8]. Completely differing from the common particle morphology, these line features marked by black arrow imply that the kappa phases in this region have been fragmented and dissolved during deformation. The dislocation slip might destroy the kappa phases and drag the solutes along with its movement [8]. Considering the affinity between segregated solutes and crystalline defects. It is highly possible that the solute segregation zones are dislocation lines. The reason why such an enrichment only takes place for Al and C, not Mn, is that the kappa phases are enriched with Al and C, not Mn [8].

5.2.2 Influences of Precipitate Characteristics on Strengthening

Precipitates can be characterized by their size, shape and the nature of the interface between a precipitate and its matrix. These factors often determine whether or not

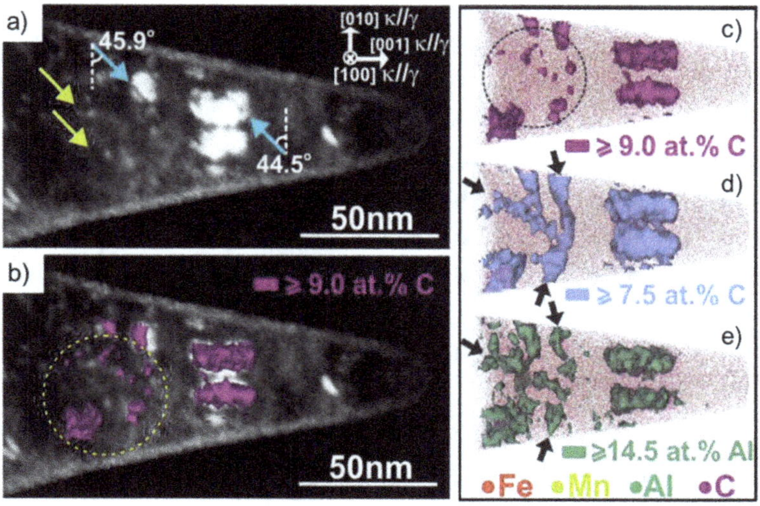

Fig. 5.7 Correlative TEM and APT analysis of a deformed microstructure at a true strain of $\varepsilon = 0.15$: **a** TEM dark-field (DF) micrograph along [001] zone axis with the blue arrows highlighting the shearing of kappa phases and yellow ones indicating the fragments of kappa phases; **b** An overlay of reconstructed APT analysis volume with purple kappa phase visualized by iso-surface C = 9.0 at.% on top of the TEM micrograph; **c–e** 3D elemental atom maps different elemental iso-surfaces of **c** C $\geq$ 9.0 at.%; **d** C $\geq$ 7.5 at.% and **e** Al $\geq$ 14.5 at.%, showing the dissolution of kappa phases [8]

a glide dislocation moving in the matrix can pass through the precipitate which will affect the mechanical properties that the material will achieve. A fully coherent interface is such that the plane of particles constituting the interface is common to the crystal structure of the particle and the matrix. This situation can only arise when the atomic spacing of the two structures in the interfacial plane are closely similar: any misfit is accommodated by elastic strain, as if shown in Fig. 5.8a. A quasi-coherent interface may be defined as one in which the regions of misfit have the characteristics of dislocations lying in the interface, as shown in Fig. 5.8b). And finally, a non-coherent interface has no special relationship between the two crystals since the dislocations have become so closely spaced that their individuality is lost [9–11]. The precipitates can be divided into: coherent precipitates, partially coherent precipitates and non-coherent precipitates. A completely coherent precipitate is one in which all its interfaces with the matrix are fully coherent, a partially coherent precipitate is one which at least one interface is coherent, and a non-coherent precipitate is one in which none of its interfaces is coherent [9–11]. An example for each type of precipitate is shown in Fig. 5.9 [11].

There is a critical dispersion condition at which the strengthening is maximum. For the deformable particles, when the dislocations pass through the particle, the intrinsic properties of the particles are of great importance. For the non-deformable particles, when the dislocations bypass the particles, the strength of the alloy is independent of the particle properties but is strongly dependent on the size of the particle the strength

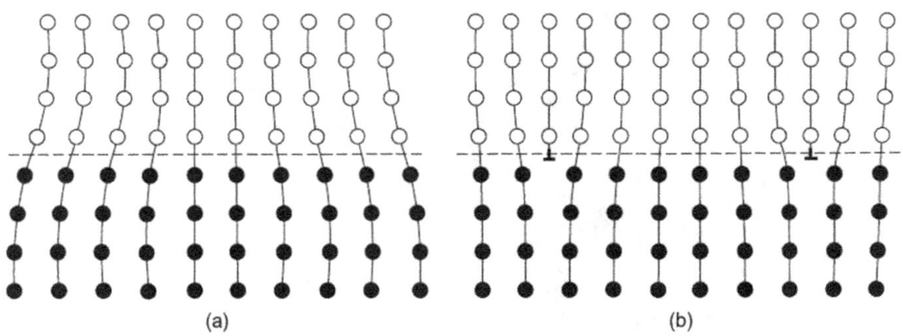

Fig. 5.8 Schematic representation of interface structures: **a** A coherent interface between two crystals of slightly different atomic spacing **b** A quasi-coherent boundary with misfit dislocations [10]

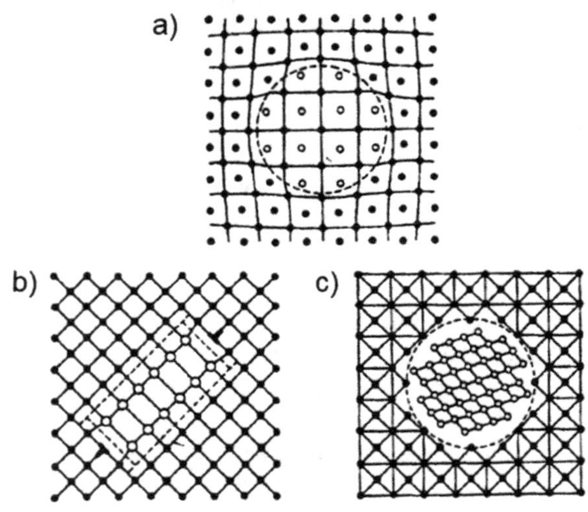

Fig. 5.9 **a** Fully coherent precipitate showing coherency strains in matrix; **b** Partially coherent precipitate showing interface dislocations; **c** Non-coherent precipitate [11]

decreases as the particle size increase [12]. Depending on the size and whether the particle is coherent, the dislocation act in different ways around the particle. In this way, the dislocation can pass through or go around (Orowan mechanism) the precipitates. The atomic structures of the microstructure with aged precipitates were characterized by Jiang et al. using scanning transmission electron microscopy (STEM) and APT techniques (Fig. 5.10) in maraging steels with a composition of Fe-18Ni-3Al-4Mo-0.8Nb-0.08C-0.01B (wt.%) [13]. The samples were annealed at 950 °C for 15 min and then aged at 500 °C for 3 h. The fast Fourier transform patterns of the image (Fig. 5.10a) taken from the < 100 > zone axis show that the matrix Is still bcc-structured, and the precipitates have an ordered B2 structure. However, no sharp particle/matrix interfaces are observed, that is, the precipitates seem to be highly coherent with the matrix. The zoom-in image of a precipitate (Fig. 5.10b) along the < 100 > direction shows the atomic columns with periodically strong and weak intensity, corresponding to the B2 ordering [13].

It is important to clarify the concepts in order to understand the theory behind the coarsening of the particles. The particles precipitate from a supersaturated solid solution to the matrix. This process is divided into three stages: nucleation, growth and coarsening. Nucleation of new phases occurs by random formation of supercritical clusters of atoms, the growth of these clusters is by diffusional transportation of atoms, and coarsening involves the growth of large particles at the expense of the dissolution of small ones, driven by an overall reduction in the interfacial energy [14]. During precipitation hardening process, the precipitates go through these stages. The evolution of the particles can be followed in an ageing diagram. It is also important to mention that the surface energy and strain energy of the precipitate with the matrix change with the precipitation process. The system follows a sequence in order to have the lowest free energy in all stages of precipitation [12].

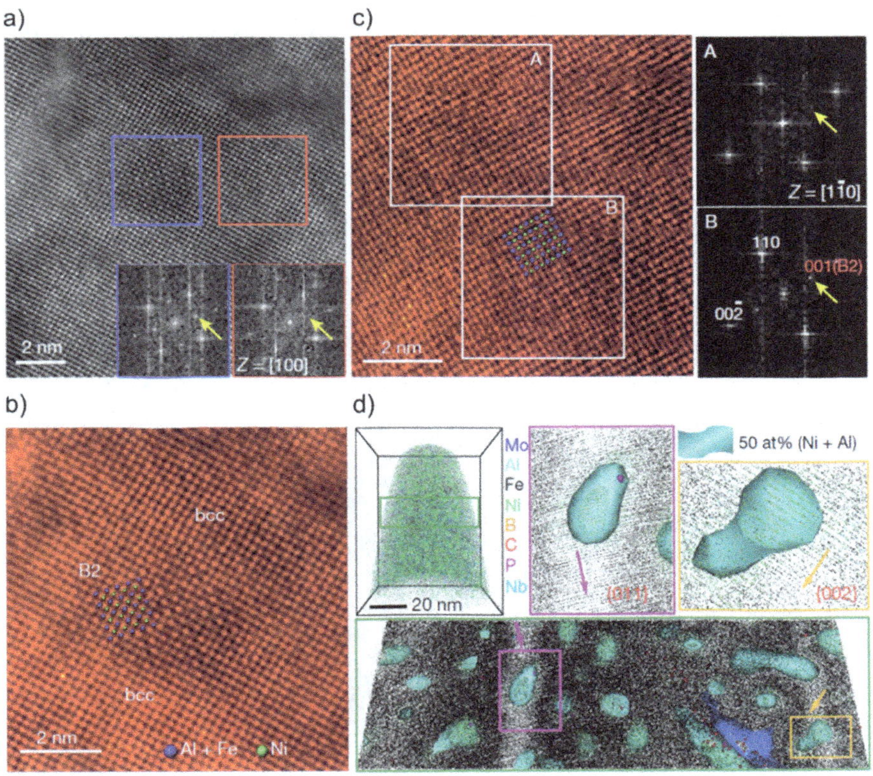

Fig. 5.10 High-angle annular dark-field STEM images and 3D APT analysis confirming the full-coherent B2 precipitates: **a** STEM image revealing the B2 ordering; **b** Zoom-in image of **a** showing the periodic atomic columns; **c** Atomic image suggesting B2 ordering; **d** The APT database showing two families of atomic planes {011} and {002} [13]

The phase-field model is used to study the morphological evolution and kinetics behaviour of precipitation. The shape of precipitates can be changed by the process parameters like precipitate size, temperature, aging time, volume fraction and cooling rate. C. Liu et al., obtained a remarkable morphology change of the γ' phase in a Ni-Al alloy due to the increment in volume fraction as is shown in Fig. 5.11 [14]. In the figure it can be seen that the γ' phase (precipitates) changes from the dispersed square particles (Fig. 5.11a) to the interconnected strip shape (Fig. 5.11d-f). It can be seen in this figure, that as the volume fraction increases, the average particles distances decrease, which indicates that the particles number density increases with the volume fraction [14].

Higher density of precipitates inhibits the dislocation motion across the grains and the distribution (homogeneous or heterogeneous) affects the dislocation motion as well. Thus, these last three parameters have a strong effect in the dislocation pinning. The pinning effect of nano-precipitates on the dislocation motion provides the key contribution to mechanical improvements such as enhancing surface hardness, stabilizing residual stress, and preventing crack initiation [15]. Furthermore, it is found that dislocation pinning significantly increases the dislocation accumulation capacity during plastic deformation [16]. Interphase precipitation is the phenomenon such that during the isothermal transformation of a steel containing strong carbide-forming elements like Ti, Nb, V, Mo and Cr, from austenite into ferrite, the alloy

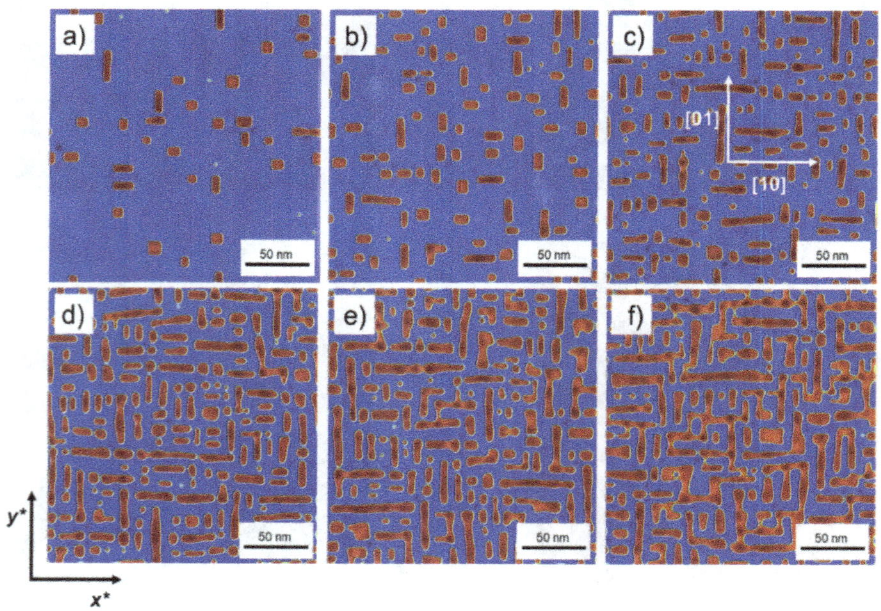

Fig. 5.11 Morphological evolution of γ' phase with increased volume fraction at $t^* = 500$. **a** $V_f = 5.2\%$; **b** $V_f = 12.9\%$; **c** $V_f = 21.3\%$; **d** $V_f = 31.6\%$; **e** $V_f = 34.2\%$; **f** $V_f = 40.2\%$. V_f = Volume fraction [14]

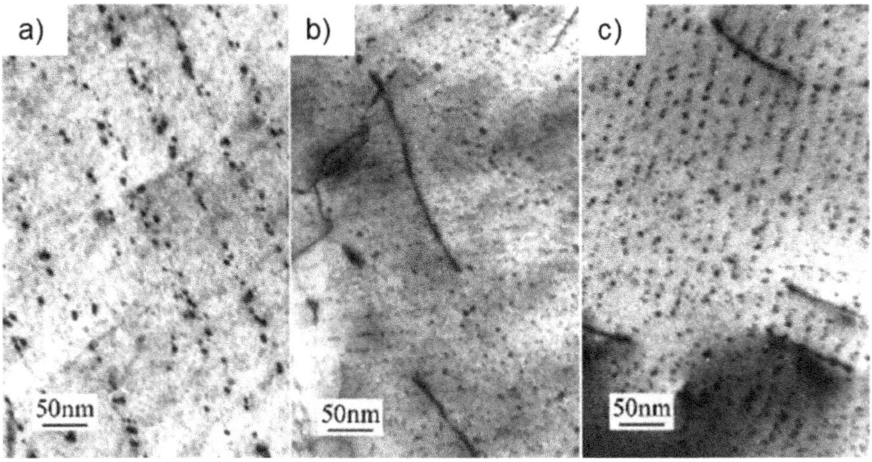

Fig. 5.12 Interphase precipitation in rolled hot strip alloys, **a** Fe-0.07C-0.68Mn-0.01Si-0.04Nb, **b** Fe-0.12C-1.21Mn-0.33Si-0.06Nb, **c** Fe-0.11C-1.54Mn-0.28Si-0.04Nb [19]

carbides precipitate on the α/γ interphase boundaries, in parallel rows as a result of periodic nucleation at migrating ferrite/austenite interfaces [17–19].

Figure 5.12 shows a TEM image of carbides form by interphase precipitation in Nb alloyed steels. Three commercial hot strip steels containing different additions of niobium and titanium were investigated [19], Fe-0.07C-0.68Mn-0.01Si-0.04Nb, Fe-0.12C-1.21Mn-0.33Si-0.06Nb, Fe-0.11C-1.54Mn-0.28Si-0.04Nb respectively. The precipitates disperse in the solid metallic matrix provides effective obstacles for dislocation motion, thus the alloys strengthened by interphase precipitation are found to have excellent mechanical properties such as high strength with reasonable elongation, high hole expansion ratio, creep resistance and good thermal stability [17, 20]. The precipitates in the matrix during the interphase precipitation can be arranged in a random distribution or well defined lines [18], but as a rule, interphase precipitation occurs in some but not in all the ferrite grains, thus this limits the contribution to the overall strength in comparison with other strengthening mechanism (*i.e.* solid solution strengthening) [19].

5.3 Nano-Precipitation Strengthening in Steels

Several nano-precipitates are assumed to be important, due to their significant contribution to yield strength, *e.g.* nano-sized carbides, nano-sized intermetallics, nano-sized ordering etc. Fe-Mn-Al-C austenitic steels with superior strain hardening behaviour have gained much research interest in the last decades [21–23]. During plastic deformation, the materials develop extraordinary service properties by the deformation mechanisms with high strain hardening potential, *i.e.* transformation-induced plasticity (TRIP) effect, twinning-induced plasticity (TWIP) effect and microband-induced plasticity (MBIP) effect [24–26]. With the addition of Al, the stacking fault energy (SFE) of the alloy increases and the Al addition promotes the precipitation of the nano-sized kappa phase $(Fe, Mn)_3AlC$ from the austenite matrix in the Fe-Mn-Al-C austenitic alloys [27–29]. The microstructure and the size of the kappa phase are strongly influenced by the aging time and aging temperature, as well as the actual alloy composition [30, 31]. Hale and Baker [30] reported the mechanical properties of the Fe-30.9Mn-8Al-1C alloy under different aging states. It was indicated that both the ultimate tensile strength and the yield strength of the steel increases with the increase of the aging temperature and reaches the maximum (ultimate tensile strength (UTS) = 1040 MPa; yield strength (YS) = 880 MPa) at 550 °C. Springer and Raabe [31] further indicated the influence of Al concentration on the microstructure and mechanical properties of the Fe-30Mn-1.2C-*x*Al alloys. High amounts of Al addition (>8 wt.%) resulted in pronounced strengthening, while low Al concentration (<2 wt.%) leads to the embrittlement of the steel during aging [31].

Intermetallic phases exhibit particular physical and mechanical properties, *e.g.* good high-temperature strength, resistance to oxidation and corrosion, low material density and high melting point, etc., and have been intensively investigated in the last decades. Earlier developments, especially in the field of maraging steels, show that intermetallic phases can be used for strengthening the hot working tools. A synergistic effect of intermetallic precipitates in combination with carbides is observed. The specific weight of car body steels can be reduced by high alloying contents of Al, that is a reduction of density by 1.3% per 1 wt.% Al addition. But, high Al alloying causes a reduction of ductility during cold rolling which has limited the Al addition to 6.5 wt.% so far; for higher Al addition the pronounced formation of the intermetallic phase Fe_3Al have been reported at annealing temperatures below 500 °C, that is below the recrystallization temperature [32, 33]. The Al-alloyed steels show outstanding properties especially at high temperatures, owing to the strengthening by the intermetallic phases in steels. It is observed that Fe_3Al with $D0_3$ structure offers more strength to the alloy when compared to FeAl with B2 structure, therefore, one of the strengthening mechanisms for Fe-Al bases alloys is stabilization of Fe_3Al at a higher temperature with respect to FeAl. The addition of alloying elements such as Ti in Fe-Al alloys can increase the critical transition temperature for the Fe_3Al/FeAl transformation. Strengthening in Fe-Al can also be achieved by precipitation hardening. Addition of alloy elements which show incomplete or partial solubility in Fe-Al based alloy introduces the possibility of strengthening these alloys by precipitation of another intermetallic compound. In several Fe-Al alloy systems such as Fe-Al-Zr, Fe-Al-Nb and Fe-Al-Ta, this intermetallic phase is an incoherent Laves phase. These incoherent precipitates provide the hindrance to the motion of dislocations and hence which increases the strength of the alloy.

Recently, a breakthrough of intermetallic control example was reported [34] by Kim et al. that by effective control of the intermetallic compound B2, the mechanical property of the investigated high strength high ductility low density steel was largely

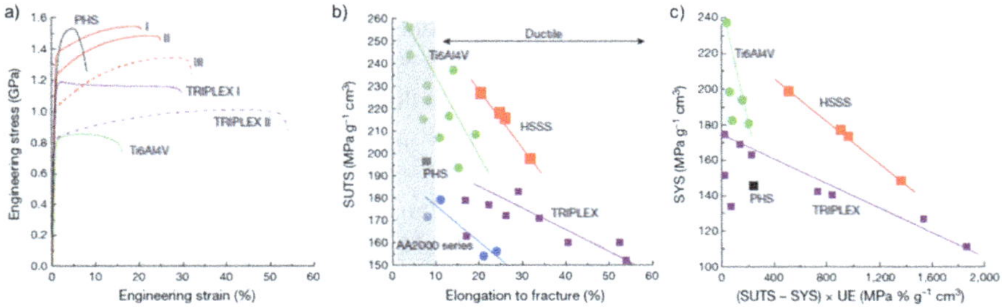

Fig. 5.13 Tensile properties of HSSS compared with selected metallic alloys of high specific strength: **a** Engineering flow curves of the annealed HSSS compared with kappa phase-strengthened TRIPLEX steels; **b** Specific ultimate tensile strength (SUTS) versus elongation of selected materials; **c** Specific yield strength (SYS) versus density-compensated tensile toughness increase during uniform elongation (UE) of HSSS compared with selected metallic alloys of high specific strength [34]

improved. By adjusting the microstructure and size distribution of the intermetallic B2 structure, the ultimate tensile strength of the steel can achieve up to 1.5 GPa. Furthermore, the large ductility and phenomenally high strain hardening capability is shown even at ultra-high yield strength levels of over 1 GPa. The mechanical properties of the investigated high-specific strength steels (HSSS) and selected metallic alloys of high specific strength are shown in Fig. 5.13.

Medium Mn steel as a candidate in the advanced lightweight material class attracts widespread attention for the combination of high strength ($\geq$1 GPa) and good ductility ($\geq$ 30%) [35–41]. Although the excellent mechanical properties of medium Mn steels are obtained, further increase in the mechanical properties of these steels by microalloying element addition achieve more and more attention [37–40]. He et al. [37] demonstrated that VC precipitates played a role in increasing the YS of ferrite by hindering the dislocation motion. Han et al. [38] reported that TiC precipitation in a 0.15 wt.% Ti-added 5 wt.% Mn steel suppressed the grain growth of austenite. Yan et al. [39] and Cheng et al. [40] designed the medium Mn steels micro-alloyed with Cu in order to control the reversed austenite, *i.e.*, the amount and stability, and improve the mechanical properties. Meanwhile, complex precipitation sequence of Cu-rich particles may remarkably influence the strengthening effect. The crystal structure evolution is usually accompanied by a change in chemical composition of Cu precipitates. In Fe-Cu-Ni-Mn alloys, the interfacial segregation of Mn and Ni at the Cu-rich particle/matrix interfaces is often observed. The elemental segregation at the interface restricts the Cu precipitates from further coarsening, leading to a core–shell structure of nano-size precipitation hardening effect. It is pointed out the C element in austenite can be decreased with the strong carbides-forming element (Nb, V, Ti) addition, which will affect the evolution of microstructure [41–44]. This is of particular importance in the context of microalloyed TRIP-assisted steels due to the interaction between strong microalloy carbide and carbonitride formers, and the evolution of microstructure [41, 44].

In addition, microalloying solute behaviour and its precipitates have a significant effect on the microstructure and mechanical properties of TRIP steels through precipitation, phase transformation and grain refinement [41–44]. Recently, attentions to the dynamic precipitation and precipitate coarsening behaviours of a combination of Mo and other microalloying elements have been raised in alloy design of microalloyed steels [41, 45–47]. The carbon activity in austenite can be decreased with the addition of Mo, which also can increase the solubility of microalloying elements in austenite and facilitate the formation of finer precipitates in the ferrite temperature range [41, 44, 45]. The misfit strain of precipitates in ferrite can be decreased through Mo incorporation, thus reducing energetic barriers at the early stages of nucleation [44, 48]. Additionally, the transformation kinetics of precipitates, ferrite and austenite can be modified by the addition of Mo [48]. Cai et al. [48] reported that the YS and the ductility of the 6.5 wt.% Mn steel can be enhanced with the addition of 0.22 wt.% Mo and 0.05 wt.% Nb. The addition of 0.19 wt.% Mo and 0.11 wt.% Ti in a 4.1 wt.% Mn steel was reported to improve the strength without loss of elongation [44, 49]. However, compared with the investigation on the single microalloy addition during thermomechanical processing, few works have been reported on the precipitation behaviours of multiple microalloy additions. Moreover, the precipitation kinetics of the multiple microalloy additions is greatly different compared with that of the single microalloy addition [41, 45–47].

By additional micro-alloying with Ce, higher Al additions up to 10.5 wt.% are applied for manufacturing cold rolled batch annealed sheet steel with a density of 6.8 g/cm^3, a yield and tensile strength of 530 MPa and 630 MPa, respectively, while the total elongation is 18% and the r-value of 1.24 [50]. This combination of mechanical properties with a low density makes this material attractive for automotive application [50]. The beneficial properties become obvious when specific values, that is yield strength range and young's modulus divided by density, are compared with alternative sheet materials like microalloyed steel HX340LAD and dual phase steels HCT780X and HCT980X, Al alloy AA5182 [50]. The specific impact of Ce alloying in a high Al containing steel is not well understood so far; it can be attributed to a formability improvement of the Fe_3Al phase [51], a reduced H intake by surface modification [52] or a texture effect [50].

5.4 Applications

5.4.1 Coherent Precipitation Strengthening in HMnS Steels

In this work, the kappa phase formation and the associated ordering effect in austenitic Fe–Mn-Al-C steels were investigated using theoretical ab initio calculations in combination with *in-situ* synchrotron X-ray diffraction methods [53]. Density function theory (DFT) calculations were performed to understand the driving forces for short-range and long-range ordering in the steels and to further assess the thermodynamic stability of the kappa phase with ordered and disordered structures. The Rietveld refinement method was used to extract the microstructural information of kappa phase strengthened Fe-Mn-Al-C steel. The contribution of the coherent kappa phase precipitation to the mechanical property will be further discussed [53].

To understand the chemistry at the atomic level in Fe-Mn-Al-C systems, the complexity is reduced by simplifying the problem: It is known that the elements Fe and Mn behave very similarly in the investigated steels compared to Al or C [54]. In order to avoid an extremely large number of different possible interactions, the calculations initially focused on an Fe-Al-C system. The numerical results were then checked with a Fe-Mn-Al-C system to verify the reliability of the simplification. However, the smaller influence of different Fe/Mn configurations on this system should be the subject of a subsequent investigation of a real quaternary system [53]. Figure 5.14a displays the calculated unit cell for carbon incorporation energies as a function of the Al-C distance [53]. For the short-range ordering, Al-Al bond was first investigated in single-layer antiferromagnetic fcc $Fe_{106}Al_2$ model cells by calculating the energy as a function of the interatomic distance. The ground magnetic state of these steels is anti-ferromagnetism in the $< 001 >$ direction, which means that all atoms of one z-layer have the same spin direction, while all atoms of the next layer have opposite directions [53].

Figure 5.15 clearly indicates repulsive Al-Al interactions in the nearest-neighboured (NN) position (labelled 1) [53]. In the antiferromagnetic (afm) model one has to distinguish between two Al atoms within one magnetic layer (↑) or in two layers with opposite spin (↑↓), and the larger value refers to the former scenario. Here, the distances are shorter (Fe afm structures are stretched tetragonally) and the magnetic afm structure is more perturbed, since both substituted Fe atoms adopt identical spin directions, resulting in a moment [53]. For 2nd NN, the same effect is present while for 3rd NN the interaction of Al atoms in neighbouring layers is

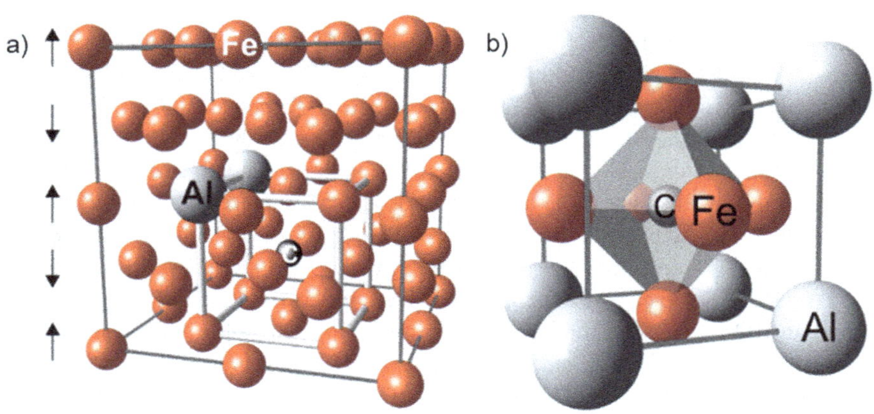

Fig. 5.14 **a** Part of a calculated unit cell for carbon incorporation energies as a function of the Al-C distance; **b** Stoichiometric kappa phase [53]

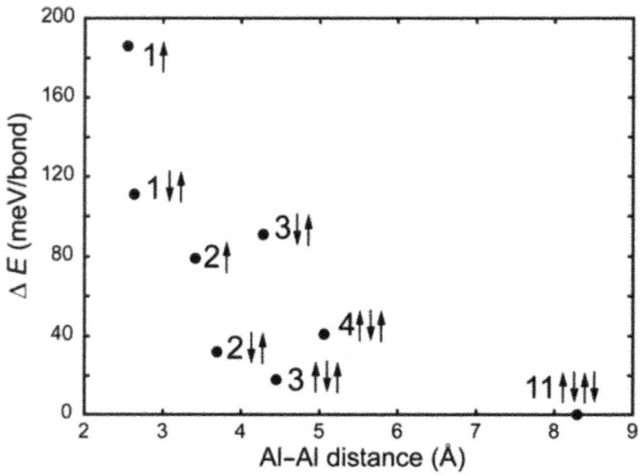

Fig. 5.15 Al-Al interaction energy as a function of interatomic distance. Numbers at the data points indicate xth (1, 2, ⋯ 11) coordination sphere while arrows indicate the number of magnetic layers between both Al atoms [53]

more repulsive (↑↓) than in second-nearest layers (↑↓↑; Al atoms are situated in two different spin-up layers with one spin-down layer in-between). Two effects are visible: The repulsive interactions increase with decreasing distance and decreasing number of interlayers. However, it must be noted that the energy differences are small and have some spread. The largest possible distance of 8.3 Å (11th NN) was assumed as the zero-interaction energy, and the maximum destabilization for direct Al-Al neighbours within a layer is about 190 meV [53]. Previous calculations [55] based on nonmagnetic 32-atom cells gave a comparable repulsion of 140/90 meV for first/second neighbours [53].

From a thermodynamic point of view, it is therefore expected that the number of direct Al-Al bonds in fcc Mn, Fe and thus also in Fe–Mn-Al-C steels is small. It shows the minimum energy configuration (see Fig. 5.16) which contains the same Al-C-Al arrangement found in kappa phases [53]. Arrows indicate the spin direction of the atoms in each layer. Figure 5.14b displays the stoichiometric of the long-range ordered kappa phase. Carbon is subsequently introduced into different octahedral sites of a solid metal lattice fcc $Fe_{106}Al_2$ with an Al-Al spacing typical of kappa phases. This allows coordination numbers (CN) of 2, 1, and 0 for carbon in C-cantered octahedral $CAl_{CN}Fe_{6\text{-}CN}$, while for CN = 0 different octahedral sites are possible, distinguished by the shortest distance to one of the two Al atoms [53]. While standard studies only consider different CNs, it is claimed that the energy is lowest at an Al-C distance, which corresponds to the structure of the kappa phase. For the longest C-Al distances, the energy is set to zero. The most favourable sites for carbon are all Al-free octahedral, and the energy increases strongly with increasing CN. For CN = 1, the energy is approx. 300 meV higher, and for CN = 2 it is even more than 1000 meV. Note that this effect is much larger than for Al-Al interactions shown in Fig. 5.15 [53]. The minimum energy site for carbon is the second-nearest neighbour shell at a distance of about 3 Å to both Al atoms, which corresponds to the configuration in the kappa phase, as stated previously. For larger Al-C distances, the interaction is constant and approx. 80 meV higher in energy [53]. It is shown that a

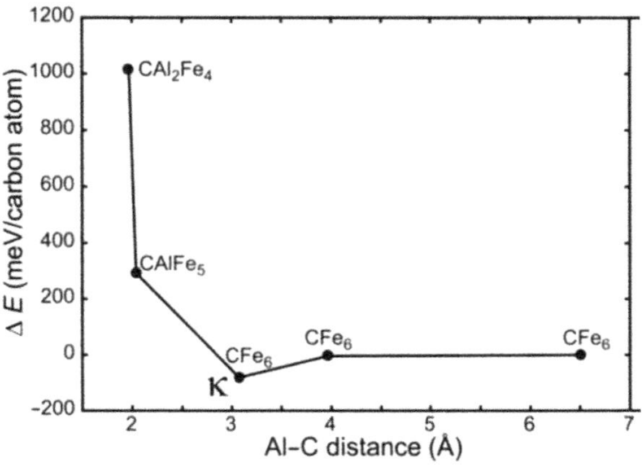

Fig. 5.16 Carbon incorporation energy differences ΔE in $Fe_{30}Al_2C$ as a function of the (closest) Al-C distance. Labels next to the data points indicate the composition of the C-surrounding octahedral. The minimum refers to a single Al-C-Al configuration which is similar to the arrangement in kappa phase. Δ E for the widest possible distance is set to 0 eV [53]

kappa-like configuration of only three atoms (a very short-range order) is apparently favoured over any other configuration in an iron matrix. On the other hand, it is known that kappa phases, which are long-ordered compounds, have very different compositions with respect to Al and C. This raises the question of the minimum Al and C concentrations considered necessary to stabilize the kappa phase [53].

Both the Al and the C content can experimentally deviate greatly from stoichiometric values [56]. To determine the lowest Al concentration, the formation energies of different compositions $Fe_{4\text{-}x}Al_x$ for the ordered and the disordered structure were calculated (Fig. 5.17a) [53]. In the latter, both types of atoms are randomly distributed on the places (as in solid solutions). In ordered structures, the x Al atoms exclusively occupy the Al sites of kappa phases (*i.e.* the corner sites of a small unit cell, as shown in Fig. 5.14) [53].

Remaining corner positions are occupied by iron, as in Al-deficient kappa phases. Only for $x = 1$ are all Al sites actually occupied by Al. Starting from $Fe_{3.75}Al_{0.25}$, where disordered and kappa configuration have very similar energies, both curves diverge with increasing Al content. Al destabilizes the random variant but stabilizes

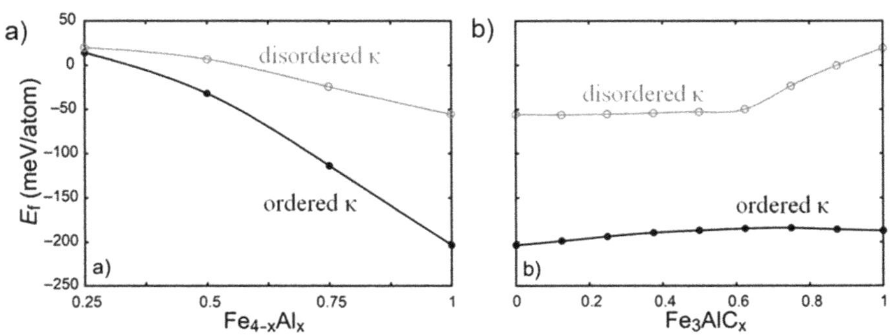

Fig. 5.17 Comparison of the formation energies between **a** Random and (partially) ordered $L1_2$ or **b** Kappa phases [53]

the ordered variant. For the stoichiometric composition Fe_3Al, it is a staggering 150 meV at 0 K. It can be assumed that at ambient temperature it is mainly the configurational entropy ΔE that changes (possibly magnetic contributions also play a minor role), which reduces the difference by about 0.05 meV/K [53]. The ordered metallic phase, called Cu_3Au type or $L1_2$, is apparently stable up to 2000 K even without carbon. The gradual addition of carbon to Fe_3Al, as shown in Fig. 17b), does not change the energy discrepancy between ordered and disordered structures up to a space occupancy of about 60% (corresponding to 3.7 wt% carbon) [53]. At higher values, the random alloy is obviously oversaturated and formation energies increase. Surprisingly, the configuration of the metals seems to have a much greater stabilizing effect on the kappa phase than carbon [53].

In combination with the theoretical calculations, in-situ SYXRD was performed to gain a deeper understanding of the kappa phase formation features at the early stage during aging [53]. During the synchrotron X-ray diffraction test, the initial microstructure in the as-quenched sample of the Fe-30Mn-8Al-1.2C alloy examined is full austenite. In the quenched condition, little kappa phase is present in the steel. The 1D synchrotron X-ray diffraction pattern of austenite (200) in the as-quenched sample of the investigated Fe-30Mn-8Al-1.2C alloy is shown in Fig. 5.18a [53]. The *in-situ* synchrotron diffraction peaks of κ (200) and γ (200) of Fe-30Mn-8Al-1.2C alloy annealed at 600 °C for 15 min and 1 h is shown in Fig. 5.18b. It turns out that the kappa phase already forms within the first hour of aging. The kappa phase diffraction peaks appear in the (200) plane at a very early stage of kappa phase formation [53]. The long-range ordered kappa phase begins to separate out of the austenitic matrix after just 15 min during aging at 600 C in Fe-30Mn-8Al-1.2C steels [53].

The in-situ SYXRD diffraction pattern in Fig. 5.19a shows the precipitation of the kappa phase from the austenitic matrix in a Fe-30Mn-8Al-1.2C alloy during aging at 600 °C for different times [53]. With increasing aging time, the intensity of the (200) peak decreases significantly, as shown in Fig. 5.19a. The diffraction peak moves slightly to the higher angle, indicating that the austenite lattice parameter is decreasing [53]. The lattice misfit value, which is defined by the formula $(a_\kappa - a_{FCC})/a_\kappa$ (where a_κ denotes the lattice constant of kappa phase and a_{fcc} is the lattice

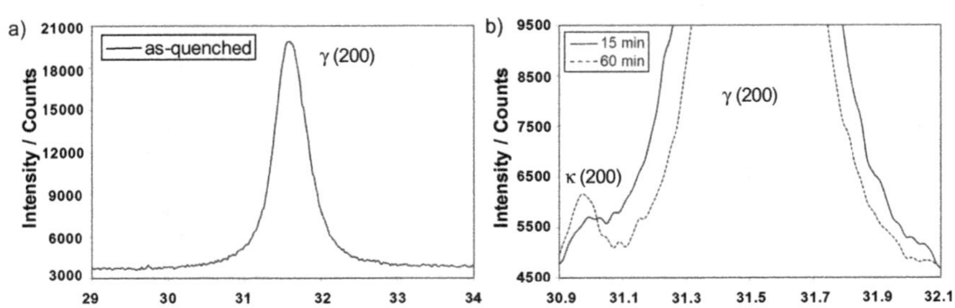

Fig. 5.18 In-situ synchrotron diffraction peaks of κ (200) and γ (200) of Fe-30Mn-8Al-1.2C alloy at the **a** after quenching at 1200 °C; and **b** annealed at 600 °C for 15 min and 1 h [53]

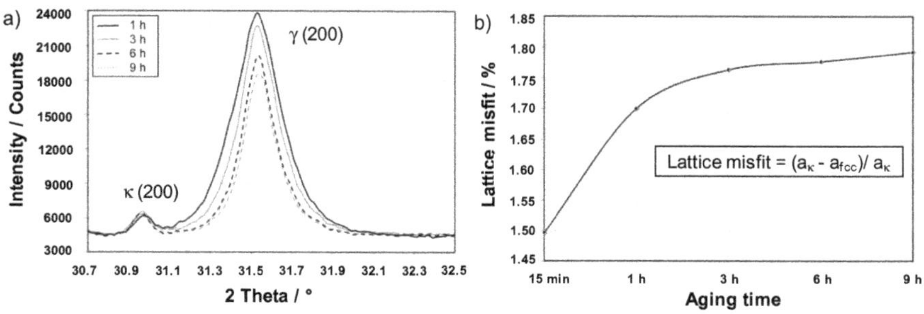

Fig. 5.19 **a** *In-situ* synchrotron diffraction peaks of κ (200) and γ (200) of Fe-30Mn-8Al-1.2C alloy annealed at 600 °C for different times; **b** The lattice misfit between the kappa phase and the austenite matrix in Fe-30Mn-8Al-1.2C alloy [53]

parameter of austenite), is a direct parameter to compare the coherence relationship between kappa phase and austenite matrix [53]. As is indicated in Fig. 5.19b, the lattice misfit between the kappa phase and the austenite matrix is very small. It shows that the lattice misfit value increases with the increase of aging time [53]. The lattice misfit value falls in the range of 1.6%-1.8%. The lattice misfit remains very small (less than 2%) up to 9 h aging. This indicates the high coherence of the kappa phase precipitation from the austenitic matrix, which contributes significantly to the strength of the materials [53].

Figure 5.20 shows the change in strength and ductility with increasing aging time during kappa phase formation [53]. With extending aging time, the yield strength and tensile strength of the Fe-30Mn-8Al-1.2C alloy both increase. With an exposure time of 15 h, the alloy has the maximum strength, UTS = 970 MPa and YS = 789MPa. The coherent nano-sized kappa phase precipitated from the austenite matrix contributes to the enhancement of mechanical properties by strong interaction with dislocations [57]. Gutierrez-Urrutia and Raabe claimed that the dominant deformation mechanism was the Orowan circumvention of particles and the successive extension of dislocation loops facilitated by cross-slipping and kappa-phase shearing [57].

The ECO index (the product of ultimate tensile strength and uniform elongation) is used to assess mechanical properties balanced by strength and ductility. The flow curves and ECO index change with increasing aging time during kappa phase formation at 600 °C in Fe-30Mn-8Al-1,2C alloy is shown in Fig. 5.21 [53]. The material shows an excellent combination of strength and ductility, especially with a short aging time, *e.g. 15 min aging.* As the aging time at 600 °C increases, the strength of the material increases and reaches its maximum at 15 h aged condition. In the Fe-30Mn-8Al-1.2C alloy in the quenched condition (UTS = 746 MPa and YS = 438 MPa) and in the short aged condition (aged 15 min, UTS = 833 MPa and YS = 542 MPa) are the massive solute in the steel result in finer cell sizes, which in turn contribute to tensile strength [53]. As shown in Fig. 5.21b, the ECO index decreases with increasing aging time, while the Fe-30Mn-8Al-1,2C alloy maintains a good combination of strength and ductility with aging times from 3 to 15 h [53].

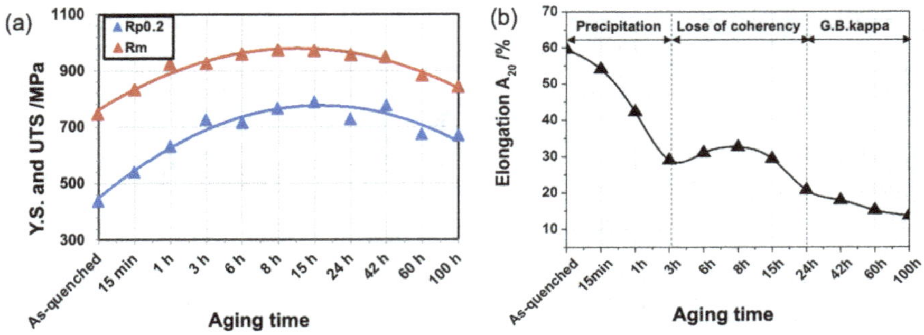

Fig. 5.20 **a** The ultimate tensile strength (UTS) and the yield strength (Y.S.) and **b** The ductility change with the increase of aging time during kappa phase formation at 600 °C in Fe-30Mn-8Al-1.2C alloy [53]

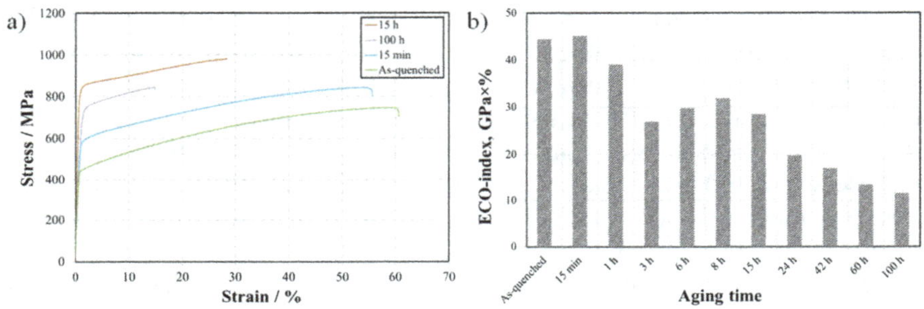

Fig. 5.21 **a** Flow curves of Fe-30Mn-8Al-1.2C steel annealed for various times; **b** The ECO-index change with the increase of aging time during kappa phase formation at 600 °C in Fe-30Mn-8Al-1.2C alloy [53]

In the overaging stage during the precipitation of the kappa phase at 600 °C, it is observed that the second population of the kappa phase with larger size at the micron level precipitates at the grain boundary, resulting in a decrease in both strength and ductility, thereby reducing the ECO index in the examined Fe-30Mn-8Al-1.2C alloy [53]. Further detailed investigations are required to investigate the influence of the elementary order and distribution, size, type and distribution of the kappa phase under different aging conditions on the deformation mechanisms and thus the work hardening behaviour of the steel [53].

5.4.2 *Microalloyed Carbides Strengthening in MMnS Steels*

In this subchapter, systematic work was carried out to study the influence of microalloying addition on the hot rolled medium Mn steels at different annealing temperatures. For analysis purposes, a comparison was made between the steels with no microalloy addition, a single Nb addition, and multiple additions of Nb-Mo to better understand the effects of 0.05 wt% Nb and 0.2 wt% Mo on the microstructure understand development and subsequent mechanical properties of the hot-rolled microalloyed TRIP-assisted steels [58]. Figure 5.22 presents the typical microstructures of the studied steels annealed at 550 °C, 600 °C, 650 °C and 700 °C for 30 min.

The microstructure of the annealed samples consists mainly of ferrite and retained austenite, which could be distinguished based on their morphological characteristics. Increasing the intercritical annealing (IA) temperature leads to a decrease in the volume fraction of ferrite. In addition, martensite is merely found in the 700 °C samples, which is transformed from austenite with low stability. It should be noted that a mixture of lamellar and equiaxed morphologies is present in all intercritically annealed specimens, which may be caused by hot rolling and the moderate strain

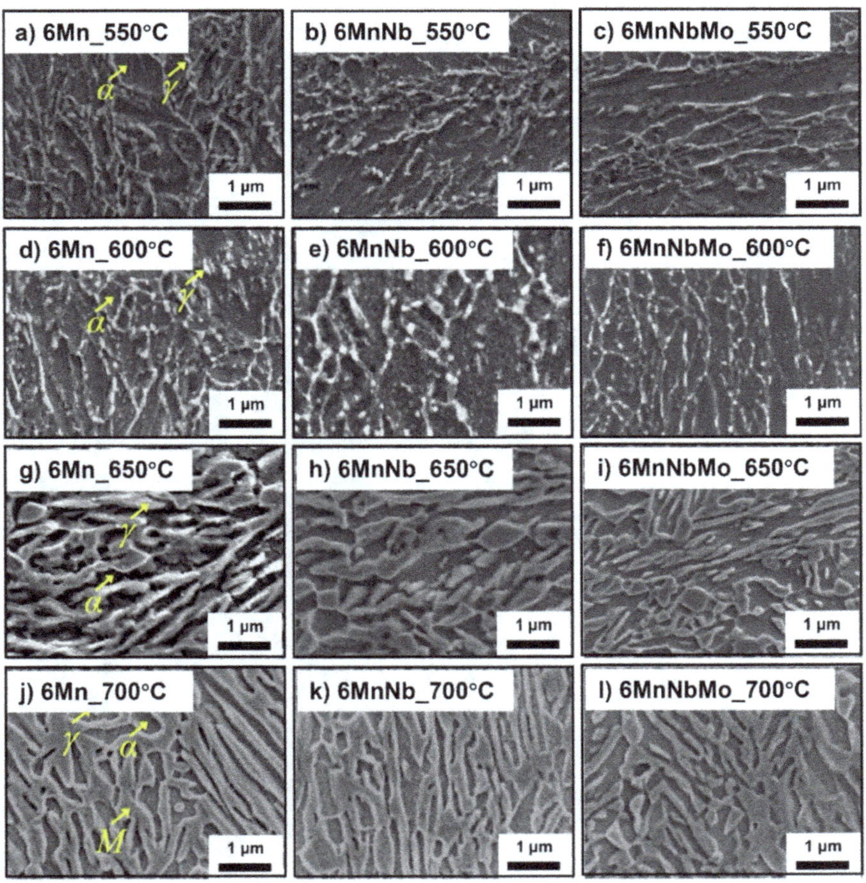

Fig. 5.22 The SEM micrographs of the warm rolled medium Mn steels 6Mn, 6MnNb and 6MnNbMo treated at 550 °C, 600 °C, 650 °C and 700 °C for 30 min; γ represents retained austenite, α represents ferrite, M represents martensite [58]

energy storage of the hot rolled specimens [39, 40]. Compared with the 6Mn sample, the addition of Nb and Mo, greatly refine the mean grain size and decreased the $V_{\gamma R}$. The TEM micrographs of the Nb(C,N) and (Nb,Mo) (C,N) at different IA temperatures are shown in Fig. 5.23 [58]. Increasing the IA temperature from 550 °C to 700 °C results in an increase in the average precipitate size of about 5 nm. Regardless of the steels containing a single Nb or multiple Nb-Mo, the precipitation density reaches the maximum at about 600 °C. In addition, the finer precipitates and higher density of precipitates can be achieved by adding Mo to the Nb-containing steels [58].

Figure 5.24a–c shows the mechanical properties of the three designed steels at different IA temperatures [58]. The tensile strengths of the three steels increase when the annealing temperature varies from 550 °C to 700 °C. Compared with the decreasing trend of the YS of the 6Mn steels from 550 °C to 700 °C, the YS of the 6MnNb and 6MnNbMo steels increase slightly from 550 °C to 600 °C, then decrease from 600 °C to 700 °C [58]. The addition of Nb or Nb-Mo decreases the total elongation when the IA temperature is lower than 650 °C, but the elongation shows an inverse trend for the higher IA temperature ($\geq$ 650 °C). In addition, compared with 6Mn steels, the Nb or Nb-Mo increases both YS and UTS. The specific YS

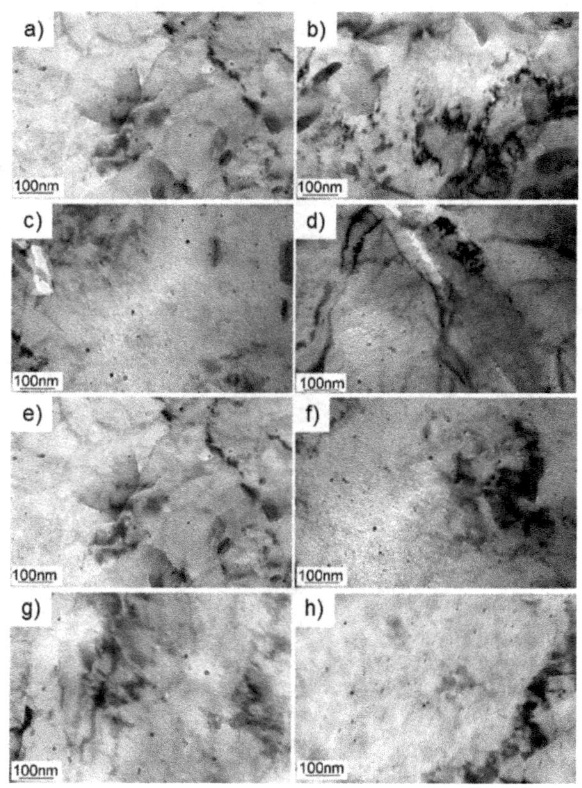

Fig. 5.23 **a, c, e g** The TEM micrographs of the precipitates of the warm rolled medium Mn steels 6MnNb and **b, d, f, h** 6MnNbMo isothermally treated at **a, b** 550 °C; **c, d** 600 °C; **e, f** 650 °C and **g, h** 700 °C for 30 min [58]

increment caused by the Nb or Nb-Mo at different IA temperatures is presented in Fig. 5.24d, and the peak performance (up to 300 MPa) is achieved at 600 °C [58].

In order to obtain the quantitative assessment of the YS contributions with addition of Nb or Nb-Mo, YS (σ_y) for analysis according to Eq. 5.3 [58]:

$$\sigma_y = \sigma_0 + \sigma_s + \sigma_g + \sigma_p + \sigma_d \tag{5.3}$$

where σ_0 denotes the ferrite friction stress (~48 MPa [59]), σ_s, σ_g, σ_p and σ_d are the individual strengthening contribution caused by solid solution, grain refinement, precipitation hardening and dislocation hardening, respectively [60]. The σ_s can be expressed using the empirical Eq. 5.4 [61]:

$$\sigma_s = 4570[C] + 4570[N] + 32[Mn] + 11[Mo] \tag{5.4}$$

where $[X_i]$ represents the mass percent of element X_i in ferrite and the value is calculated by the JmatPro. The σ_g can be calculated according to the Hall–Petch Eq. 5.5.

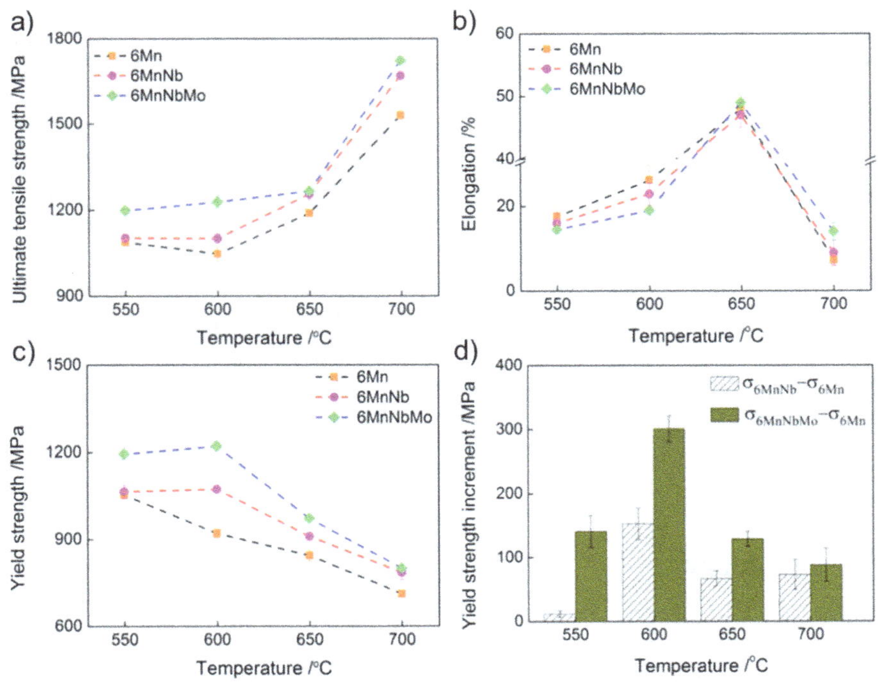

Fig. 5.24 **a** Ultimate tensile; **b** Strength elongation; **c** Yield strength and **d** The increment of yield strength of the warm rolled medium Mn steels 6Mn, 6MnNb and 6MnNbM treated at 550 °C, 600 °C, 650 °C and 700 °C for 30 min: σ_{6Mn}-yield strength of steels 6Mn, σ_{6MnNb}-yield strength of steels 6MnNb and $\sigma_{6MnNbMo}$-yield strength of steels 6MnNbMo [58]

$$\sigma_g = k_y * d^{-0.5} \tag{5.5}$$

where d is the mean grain size (Table 5.1) and k_y is a constant (~17.4 MPa/mm$^{-0.5}$). For dislocation strengthening calculation, the Bailey-Hirsch relationship is used [59]:

$$\sigma_d = M\alpha Gb\sqrt{\rho_{total}} \tag{5.6}$$

here, M is the Taylor factor (~2.75 for random textured bcc metals [62]), α is a constant (~0.38 in bcc iron [62], G is the shear modulus (~81.6 GPa), b is the Burgers vector and ρ_{total} (Table 5.1) is the total dislocation density. The precipitation hardening is given by the Ashby-Orowan relation as follows [62]:

$$\sigma_p = \frac{9f^{0.5}}{d^*} ln(1640d^*) \tag{5.7}$$

where, f is the volume fraction of the precipitates and d^* is the average precipitate diameter (Table 5.1).

The calculated values for the different strengthening effects are shown in Fig. 5.25 [58], which shows a relatively good agreement with the experimental results. It is clear that the Nb or Nb-Mo addition makes a significant contribution to the YS mainly through grain refinement and precipitation hardening. Compared with the 6Mn steel, the addition of Nb or Nb-Mo increases the YS of 285 MPa or 379 MPa through precipitation hardening and increases the YS of 16 MPa or 33 MPa due to grain refinement at 600 °C [58]. The stability of retained austenite and the precipitation behaviours of the warm rolled medium Mn steels predominate the mechanical properties. The addition of Nb or Nb-Mo increases the YS from 12 MPa—153 MPa or 88 MPa—301 MPa and increase the 15 MPa—139 MPa or 75 MPa—192 MPa UTS compared to the steels without the addition of micro-alloying elements. The maximum increment of the YS is achieved at about 600 °C. The addition of Nb or Nb-Mo can reduce the volume fraction of retained austenite but increase the mechanical stability of austenite by refining the average grain size of austenite [58]. Mo addition can increase the NbC precipitate density and decrease the average precipitate size by lowering the interfacial energy between the precipitates and the matrix [58].

5.5 Summary

With the benefits of modern experimental methods, explicit inspection and characterization of nano-sized precipitates can be obtained. Remarkable results from literature in the last decades and own-related research work revealed the strengthening effect as the function of precipitate type, size, distribution and boundary coherency. Nano-sized precipitation has been long researched and regarded as an important strengthening approach in nano-engineering.

Table 5.1 The mean grain size (d), the total dislocation density (ρ_{total}), the average precipitates size (d^*) and the precipitation volume fraction of the warm-rolled medium Mn steels 6Mn, 6MnNb and 6MnNbMo with different IA temperature [58]

Temperature/°C	$\bar{d}$/um			ρ_{total}/m^{-2}			d^*/nm			f		
	6Mn	6Mn Nb	6Mn NbMo	6Mn	6Mn Nb	6Mn NbMo	6Mn	6MnNb	6Mn NbMo	6Mn	6MnNb	6Mn NbMo
550	1.2	1.1	1	2.1×10^{14}	2.0×10^{14}	2.1×10^{14}	0	6.4	6.1	0	3.0×10^{-4}	5.1×10^{-4}
600	1.5	1.4	1.3	1.6×10^{14}	1.6×10^{14}	1.9×10^{14}	0	6.8	5.9	0	7.9×10^{-4}	12×10^{-4}
650	1.7	1.5	1.4	1.5×10^{14}	1.6×10^{14}	1.9×10^{14}	0	8.1	8	0	6.1×10^{-4}	10×10^{-4}
700	1.9	1.8	1.5	1.3×10^{13}	1.3×10^{13}	1.4×10^{13}	0	10.2	9.9	0	4.0×10^{-4}	7.1×10^{-4}

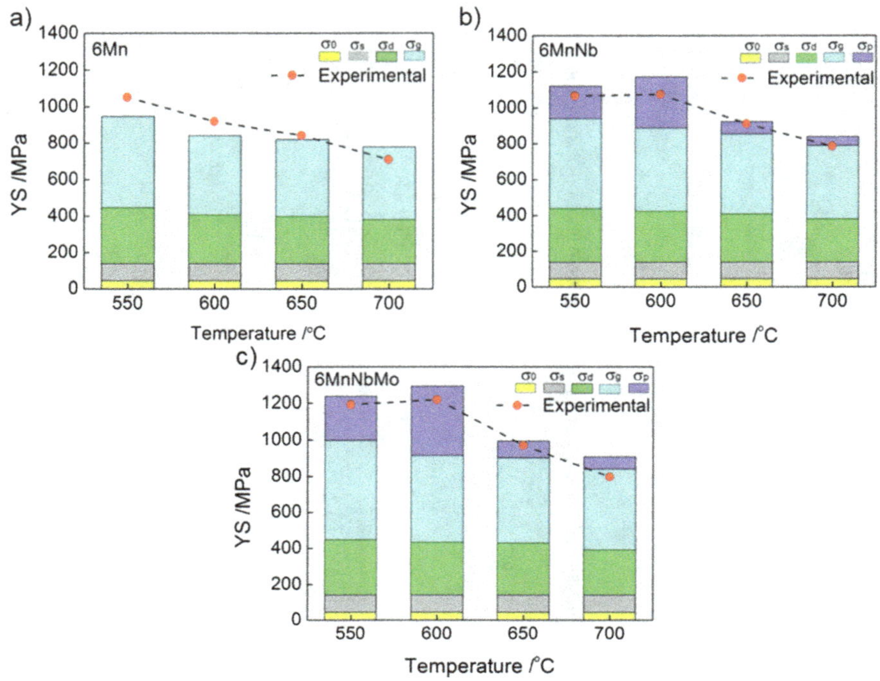

Fig. 5.25 The quantification of solid solution strengthening, grain refinement strengthening, dislocation hardening and precipitation hardening of the warm-rolled medium Mn steels **a** 6Mn, **b** 6MnNb and **c** 6MnNbMo at different IA temperature [58]

References

1. Callister W (2007) Materials science and engineering: an introduction, 7th edn. John Wiley & Sons, New York
2. Gladman T (1999) Precipitation hardening in metals. Mater Sci Technol 15:30–36
3. Groh S (2014) Transformation of shear loop into prismatic loops during bypass of an array of impenetrable particles by edge dislocations. Mater Sci Eng A 618:29–36
4. Myhr O, Grong Ø, Pedersen K (2010) A combined precipitation, yield strength, and work hardening model for Al-Mg-Si alloys. Metall Mater Trans A 41:2276–2289
5. Maciejewski K, Jouiad M, Ghonem H (2013) Dislocation/precipitate interactions in IN100 at 650 °C. Mater Sci Eng A 582:47–54
6. Smallman R, Bishop R (1999) Modern physical metallurgy and materials engineering, 6th edn. Elsevier, Oxford
7. Maciejewski K, Ghonem H (2014) Isotropic and kinematic hardening as functions of gamma prime precipitates in a nickel-based superalloy. Int J Fatigue 68:123–135
8. Yao M (2017) κ-carbide in a high-Mn light-weight steel: precipitation, off-stoichiometry and deformation, Doctoral Thesis, RWTH Aachen University
9. Martin J (1968) The basis of age-hardening, in: Precipitation hardening (first ed.), Oxford, Pergamon Pres
10. Smallman RE, Ngan AHW (2014) Chapter 10—Surfaces, Grain Boundaries and Interfaces. In Modern Physical Metallurgy (eighth ed.), Butterworth-Heinemann, 2014, 415–442.

11. Martin J (1998) The microstructure of precipitation-hardened alloys, in: Precipitation Hardening (second ed.), Oxford, Butterworth Heinemann
12. Smallman R (2014) Precipitation hardening, in: Smallman R, Ngan A (Eds.), Modern physical metallurgy, Oxford, Butterworth-Heinemann
13. Jiang S, Wang H, Wu Y, Liu X, Chen H, Yao M, Gault B, Ponge D, Raabe D, Hirata A, Chen M, Wang Y, Lu Z (2017) Ultrastrong steel via minimal lattice misfit and high-density nanoprecipitation. Nature 544:460–464
14. Liu C, Li Y, Zhu L, Shi S, Chen S, Jin S (2018) Morphology and kinetics evolution of γ' phase with increased volume fraction in Ni-Al alloys. Mater Chem Phys 217:23–30
15. Liao Y, Ye C, Gao H, Kim B, Suslov S, Stach E, Cheng G (2011) Dislocation pinning effects induced by nano-precipitates during warm laser shock peening: dislocation dynamics simulation and experiments. J Appl Phys 110:023518
16. Ye C, Liao Y, Suslov S, Lin D, Cheng G (2014) Ultrahigh dense and gradient nano-precipitates generated by warm laser shock peening for combination of high strength and ductility. Mater Sci Eng 609:195–203
17. Zhang Y, Miyamoto G, Shinbo K, Furuhara T, Ohmura T, Suzuki T, Tsuzaki K (2015) Effects of transformation temperature on VC interphase precipitation and resultant hardness in low-carbon steels. Acta Mater 84:375–384
18. Davenport A, Brossard L, Miner R (1975) Precipitation in microalloyed high strength low-alloy steels. JOM 27:21–27
19. Kestenback H, Campos S, Morales E (2006) Role of interphase precipitation in microalloyed hot strip steels. Mater Sci Technol 22:615–626
20. Seol J, Na S, Gault B, Kim J, Han JC, Park C, Raabe D (2017) Core-shell nanoparticle arrays double the strength of steel. Sci Rep 7:42547
21. Kim H, Suh D, Kim N (2013) Fe-Al-Mn-C lightweight structural alloys: a review on the microstructures and mechanical properties. Sci Technol Adv Mater 14:014205
22. Bouaziz O, Allain S, Scott C, Cugy P, Barbier D (2011) High manganese austenitic twinning-induced plasticity steels: a review of the microstructure properties relationships. Curr Opin Solid State Mater Sci 15:141–168
23. Byun T (2003) On the stress dependence of partial dislocation separation and deformation microstructure in austenitic stainless steels. Acta Mater 51:3063–3071
24. Frommeyer G, Brüx U, Neumann P (2003) Supra-ductile and high strength manganese-TRIP/TWIP steels for high energy absorption purposes. ISIJ Int 43:438–446
25. Barbier D, Gey N, Allain S, Bozzolo N, Humbert M (2009) Analysis of the tensile behavior of a TWIP steel based on the texture and microstructure evolutions. Mater Sci Eng A 500:196
26. Lee S, Estrin Y, De Cooman B (2013) Effect of the strain rate on the TRIP-TWIP transition in austenitic Fe-12 pct Mn-0.6 pct C TWIP Steel, Metall. Mater. Trans. A 45 717–730
27. Song W, Ingendahl T, Bleck W (2014) Control of strain hardening behavior in high-Mn austenitic steels. Acta Metall. Sinica (English Letters) 27:546–556
28. Gutierrez-Urrutia I, Raabe D (2013) Influence of Al content and precipitation state on the mechanical behavior of austenitic high-Mn low-density steel. Scr Mater 68:343–347
29. Chao C, Hwang C, Liu T (1993) Grain boundary precipitation in an Fe-7.8Al-31.7Mn-0.54C alloy, Scr Metall Mater 28 109–114
30. Hale G, Baker A (1986) Carbide precipitation in austenitic Fe-Mn-Al-C alloys, presented at Conf on Alt Alloying for Env Res, New Orleans
31. Springer H, Raabe D (2012) Rapid alloy prototyping: compositional and thermo-mechanical high throughput bulk combinatorial design of structural materials based on the example of 30Mn-1.2C-xAl triplex steels, Acta Mater. 60 4950–4959
32. Schwabe J (2014) Entwicklung von dichtereduzierten Feinblechstählen für Automobilanwendungen, Doctoral thesis, RWTH Aachen University
33. Ikeda O, Ohnuma I, Kainuma R, Ishida K (2001) Phase equilibria and stability of ordered bcc phases in the Fe-rich portion of the Fe-Al system. Intermetallics 9:755–761
34. Kim S, Kim H, Kim N (2015) Brittle intermetallic compound makes ultrastrong low-density steel with large ductility. Nature 518:77

35. Han J, Silva A, Ponge D, Raabe D, Lee S, Lee Y, Lee S, Hwang B (2017) The effects of prior austenite grain boundaries and microstructural morphology on the impact toughness of intercritically annealed medium Mn steel. Acta Mater 122:199–206
36. Pan H, Cai M, Ding H, Huang H, Zhu B, Wang Y, Zhang Y (2017) Microstructure evolution and enhanced performance of a novel Nb-Mo microalloyed medium Mn alloy fabricated by low-temperature rolling and warm stamping. Mater Des 134:352–360
37. He B, Hu B, Yen H, Cheng G, Wang Z, Luo H, Huang M (2017) High dislocation density-induced large ductility in deformed and partitioned steels. Science 357:1029–1032
38. Han Y, Shi J, Xu L, Cao W, Dong H (2011) TiC precipitation induced effect on microstructure and mechanical properties in low carbon medium manganese steel. Mater Sci Eng A 530:643–651
39. Yan S, Liang T, Chen J, Li T, Liu X (2019) A novel Cu-Ni added medium Mn steel: precipitation of Cu-rich particles and austenite reversed transformation occurring simultaneously during ART annealing. Mater Sci Eng A 746:73–81
40. Cheng P, Hu B, Liu S, Guo H, Enomoto M, Shang C (2019) Influence of retained austenite and Cu precipitates on the mechanical properties of a cold-rolled and intercritically annealed medium Mn steel. Mater Sci Eng A 746:41–49
41. Abbasi E, Rainforth W (2017) Microstructural evolution of Nb-V-Mo and V containing TRIP-assisted steels during thermomechanical processing. J Mater Sci Technol 33:311–320
42. Kim J, Jung J, Kim D, Lee Y (2013) The kinetics of Nb(C, N) precipitation during the isothermal austenite to ferrite transformation in a low-carbon Nb-microalloyed steel. Acta Mater 61:7437–7443
43. Kapoor M, Kaub T, Darling K, Boycec B, Thompson G (2017) An atom probe study on Nb solute partitioning and nanocrystalline grain stabilization in mechanically alloyed Cu-Nb. Acta Mater 126:564–575
44. Pan H, Ding H, Cai M (2018) Microstructural evolution and precipitation behavior of the warm-rolled medium Mn steels containing Nb or Nb-Mo during intercritical annealing. Mater Sci Eng A 736:375–382
45. Enloe C, Findley K, Parish C, Miller M, De Cooman B, Speer J (2013) Compositional evolution of microalloy carbonitrides in a Mo-bearing microalloyed steel. Scr Mater 68:55–58
46. Jiang L, Marceau R, Dorin T, Hodgson P, Stanford N (2019) Effect of molybdenum on phase transformation and microstructural evolution of strip cast steels containing niobium. J Mater Sci 54:1769–1784
47. Cao J, Yong Q, Liu Q, Sun X (2007) Precipitation of MC phase and precipitation strengthening in hot rolled Nb-Mo and Nb-Ti steels. J Mater Sci 42:10080–10084
48. Cai M, Li Z, Chao Q, Hodgson P (2014) A novel Mo and Nb microalloyed medium Mn TRIP steel with maximal ultimate strength and moderate ductility. Metall Mater Trans A 45:5624–5634
49. Lee D, Kim J, Lee S, Lee K, De Cooman B (2017) Microstructures and mechanical properties of Ti and Mo micro-alloyed medium Mn steel. Mater Sci Eng A 706:1–14
50. Bleck W, Song W, Zimmermann A (2016) Intermetallic phases in new steels. Mater Sci Forum 879:9–14
51. Sun Y, Yao Z, Zhang Z, Huang H (1995) Mechanical properties of Fe_3Al-based alloys with cerium addition. Scr Metall Mater 33:811–817
52. Stoloff N, Liu C (1994) Environmental embrittlement of iron aluminides. Intermetallics 2:75–87
53. Song W, Zhang W, von Appen J, Dronskowski R, Bleck W (2015) κ-phase formation in Fe-Mn-Al-C austenitic steels. Steel res int 86:1161–1169
54. Yen H, Ooi S, Eizadjou M, Breen A, Huang C, Bhadeshia H, Ringer S (2015) Role of stress-assisted martensite in the design of strong ultrafine-grained duplex steels. Acta Mater 82:100–114
55. von Appen J, Dronskowski R (2011) Carbon-induced ordering in manganese-rich austenite-a density-functional total-energy and chemical-bonding study. Steel Res Int 82:101–107

56. Monkhorst H, Pack J (1976) Special points for Brillouin-zone integrations. Phys Rev B 13:5188–5192
57. Gutierrez-Urrutia I, Raabe D (2014) High strength and ductile low density austenitic FeMnAlC steels: simplex and alloys strengthened by nanoscale ordered carbides. Mater Sci Tech 30:1099–1104
58. Pan H, Ding H, Cai M, Kibaroglu D, Ma Y, Song W (2019) Precipitation behaviour and austenite stability of Nb or Nb-Mo micro-alloyed warm-rolled medium-Mn steels. Mater Sci Eng A 766:138371
59. Ghosh S, Bandyopadhyay P, Kundu S, Chatterjee S (2011) Copper bearing microalloyed ultrahigh strength steel on a pilot scale: microstructure and properties. Mater Sci Eng A 528:7887–7894
60. Gottstein G (2004) Physical foundation of materials science, 1st edn. Springer, New York
61. Jang J, Lee C, Heo Y, Suh D (2012) Stability of (Ti, M)C (M = Nb, V, Mo and W) carbide in steels using first-principles calculations. Acta Mater 60:208–217
62. Wu D, Han X, Tian H, Liao B, Xiao F (2015) Microstructural characterization and mechanical properties analysis of weld metals with two Ni contents during post-weld heat treatments. Metall Mater Trans A 46:1973–1984

Chapter 6
Interface Engineering

Interfaces show significant influence on the various phase transformations during heat treatments and/or deformation of the metallic materials. This chapter describes the interface engineering approach and its application in the high strength steels. Own-related research work[1,2] of precipitation engineering is reported as application examples in Sect. 6.4.

6.1 Definition and History

In metallic materials, interfaces usually refer to the conjunction of two phases or two grains. It has been found in the last decades that the interfaces in metallic materials play a very important role in controlling the electrical, optical, magnetic and mechanical properties of the materials, in particular at nano-scale. Interfaces are considered as determining factor during the forming or different manufacturing process since the interfaces together with the grains and the other crystal structure parameters strongly influence the resulting mechanical properties, e.g. strength, formability limits, machinability and weldability. The concept of interface engineering and grain boundary engineering are initiated recently for controlling, developing and enhancing the properties of the modern materials. In 1940, C. S. Smith studied the large variety of grain structures in polycrystalline materials and the evolution of the grain structure [1]. The importance of the interfacial energy on the interface behaviour in single phase and multiple phase alloys was first-time addressed. In the 1950s, the microstructural aspects in terms of the interfaces and grain boundary behaviours were intensively

Own work-related examples link to the publications:

[1] Shen [30].

[2] Song [37].

W. Song, *Nano-Engineering of High Strength Steels*,
Topics in Mining, Metallurgy and Materials Engineering,
https://doi.org/10.1007/978-3-031-42967-5_6

discussed [2, 3]. During the 1960s–1980s, the rapid development of computer simulations and sophisticated characterization methods enabled investigating the interfaces by both simulation and experimental approaches. The focus was devoted to the grain boundary structure and its influence on material properties. Since the 1990s, high-resolution transmission electron microscopy (HRTEM) has largely contributed to the characterization of the atomic structures of grain and phase boundaries [4, 5]. In recent years, interface engineering approach has been proposed to enhance the materials properties via controlling the interface structures, chemical gradient at the interfaces, misorientation parameters, etc.

6.2 Interfaces Characteristics During Phase Transformations

The solid-state phase transformations in metallic materials can be controlled either by diffusion or by interface movement in the materials. In diffusion-controlled phase transformations, the phase transformations are controlled mainly by long-range diffusion of alloying elements [6, 7]. In interface-controlled phase transformations, the diffusion is assumed as infinite and the phase transformations are considered to be controlled by the movement of the interface, which is dependent on the rearrangement of atoms at the interface during its movement [6, 7]. The movement of the interface is generally explained by means of three different models, i.e. sharp interface, finite interface and diffuse interface models. In sharp interface model, the velocity of the interface is calculated by considering the jumps of individual atoms or the fluxes between different phases. In finite interface is based on the concentration profile throughout the interface with a finite width. In diffuse interface model, the position of the interface is determined by means of phase field parameters, which vary depending on the width of the interface [6].

The interstitial and substitutional alloying elements may alter the equilibrium conditions at the interfaces due to their different diffusivity and solubility in the parent phase and in the newly-formed phases [6]. This leads to a deviation from the ideal state of a thermodynamic equilibrium of the materials. Three equilibrium conditions are commonly used to describe the deviation, i.e. local equilibrium with partitioning (LE/PLE); none (negligible) partitioning local equilibrium mode (NPLE); para-equilibrium (PE). The local equilibrium with partitioning mode refers to the state when the equilibrium is reached at the interface by the partitioning of both interstitial and substitutional alloying elements between two phases. Figure 6.1a shows the schematic isothermal sections of a Fe-C-Mn ternary system during ferrite growth under PLE mode [8].

Figure 6.1b shows an alternative scenario to choose the tie-line, where the transformation process is controlled by only the diffusion of interstitial elements. When the equilibrium is governed only by the diffusion of interstitial atoms and there is no partitioning (or negligible partitioning) of substitutional solute atoms at the interface,

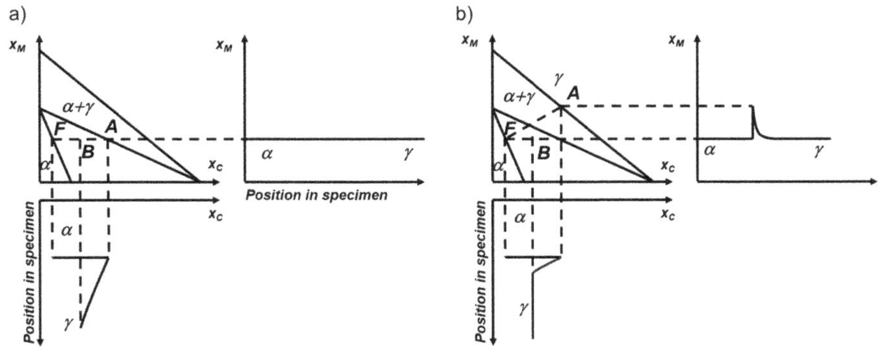

Fig. 6.1 Schematic isothermal sections of a Fe-C-Mn ternary system during ferrite growth under **a** PLE mode and **b** NPLE mode at the interface [8]

the partitioning mode is called NPLE mode. The growth rate during the transformation under NPLE mode is much higher than that in the PLE mode [9]. Under NPLE mode, a concentration spike with the depletion or enrichment of solute atoms often occurs at the interface. PE mode refers a constrained equilibrium condition which is observed when the concept of local equilibrium around the interface is no longer valid due to decreased temperatures [6, 8]. Under PE mode, the phase transformation is controlled only by the diffusion of interstitial elements [9]. The transition between these three mechanisms is usually related to the chemical composition of alloying elements and the temperature.

6.3 Interface Engineering in High Strength Steels

The interface engineering is a powerful approach to develop novel high strength steels with excellent mechanical properties. It is widely applied in high strength steels by using various types of planar defects, e.g. the prior austenite grain boundaries, twin boundaries, austenite/ferrite phase boundaries, etc. As the newly-developed 3rd generation of advanced high strength steels, the interface characteristics in medium-Mn steels considering element partitioning through austenite-ferrite and precipitate-matrix interfaces are of particular interest. Figure 6.2 shows the austenite growth kinetics depending on time and temperature during intercritical annealing of a medium-Mn Fe-0.18C-11Mn-3.8Al (wt.%) steel [10]. During intercritical annealing (IA), partitioning of C and Mn from initially existing martensite to austenite occurs. Depending on the partitioning kinetics, IA process usually contains three stages as depicted in Fig. 6.2.

In the first stage, the transformation is mainly controlled by C diffusion in ferrite under NPLE conditions and the fraction of austenite increases rapidly. In the second stage, the transformation is mainly controlled by diffusion of Mn in ferrite under PLE conditions and the austenite growth is relatively slow. In this stage, Mn is often enriched at the austenite-ferrite interface, due to the low diffusivity of Mn in austenite.

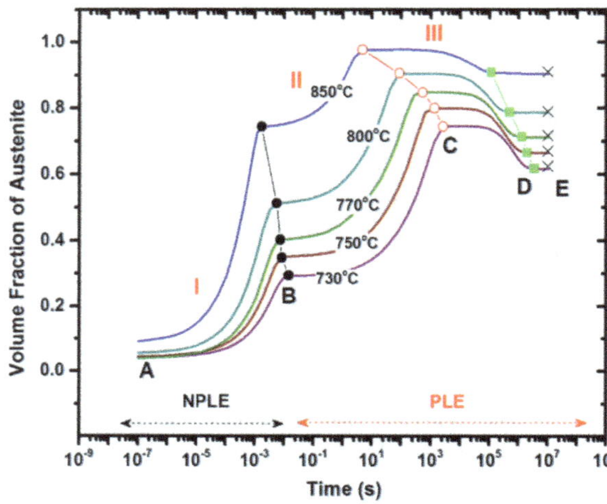

Fig. 6.2 Austenite growth kinetics depending on time and temperature during intercritical annealing of medium-Mn-Fe-0.18C-11Mn-3.8Al (wt.%) steel [10]

The enrichment continues until the interface stops moving and maximum austenite fraction is reached. Then the migration of the interface in the reverse direction starts in the third stage, also named as austenite shrinkage [9, 10]. The third stage is controlled by very slow Mn diffusion in austenite under PLE conditions. Homogenous distribution of C, Mn, and Al in austenite takes place and equilibrium volume fraction of austenite is achieved in this stage [10]. It was reported by Tirumalasetty et al. that Mn enrichment at austenite-ferrite phase boundary may reduce the mobility of the interface and therefore increases the stability of ultrafine grained microstructure due to the solute drag effect [11]. Nano-clusters and nano precipitates may also induce the pinning of the migrating austenite-ferrite interface. When the interface is stagnant, the precipitation starts in the vicinity of the interface or in the ferrite matrix [12]. Tirumalasetty et al. illustrated that the nano precipitates with reduced lattice misfit from the matrix show great strengthening effect. The reduced lattice misfit lowers interfacial energy and eases the nucleation. As a result, the coarsening rate is slowed, and the finer precipitates are obtained [11].

The effects of prior austenite grain boundaries and microstructural morphology on the impact toughness of an annealed Fe-7Mn-0.1C-0.5Si medium Mn steel were investigated by Han et al. for two different microstructure states, hot-rolled-annealed (HRA) and cold rolled-annealed (CRA) [13]. The HRA specimens reveal lath-shaped morphology and the CRA specimens have globular-shaped morphology. The HRA specimen have higher ductile to brittle transition temperature and lower low-temperature impact toughness compared to the CRA specimen. This is due to intergranular cracking in the HRA specimens along prior austenite grain boundaries decorated by C, Mn and P [13]. In the CRA specimen intergranular cracking occurs along the boundaries of the very fine ferrite and martensite grains. The results manifest that cold working prior to intercritical annealing promotes the elimination of the solute-decorated boundaries of coarse prior austenite grains through the recrystallization of martensite prior to reverse transformation, hence improving the low-temperature impact toughness of medium Mn steel [13].

To confirm the difference in the degree of segregation pertaining to a certain type of interfaces, atom probe tomography (APT) analysis is conducted on an as-hot-rolled specimen which was air-cooled to room temperature (Fig. 6.3). Figure 6.3b shows the three-dimensional APT result. C iso-surfaces with a concentration of ~2 at.% are plotted to highlight the decorated grain boundaries (GB1 and GB2). The C segregation to the boundaries is assumed to occur primarily during air cooling at temperatures of below the martensite start (M_s) temperature due to the high mobility of the C atoms [13]. Figure 6.3c shows the one-dimensional elemental profiles corresponding to interfaces GB1 and GB2. While GB1 reveals only C segregation, GB2 is decorated by C, Mn and P [13]. Both Mn and P are known as harmful elements reducing the grain boundary cohesion. As substitutional Mn and P atoms cannot diffuse sufficiently far below the M_s temperature, both elements must have segregated to the boundaries already in the single-phase austenite region at high temperatures above M_s [13]. This observation indicates that GB2 is a prior austenite grain boundary and GB1 is a boundary pertaining to martensite packets or blocks which formed at low temperatures below the M_s temperature [13].

The deep understanding of interface behaviour and elemental partitioning behaviour are achieved by the means of cyclic phase transformation [14–17]. Chen et al. [14] employed cyclic phase transformation in Fe-C-Mn system to avoid the

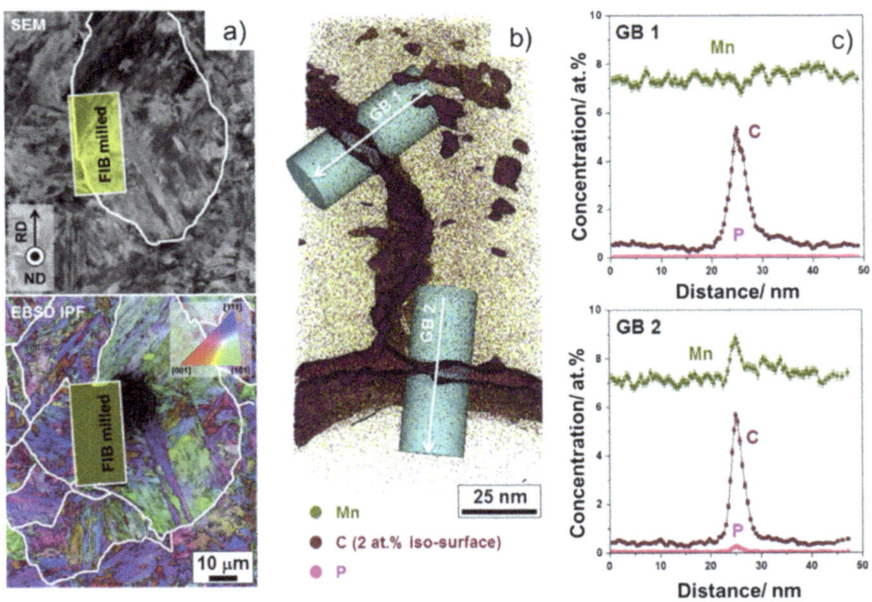

Fig. 6.3 **a** Scanning electron microscopy (SEM) and electron backscatter diffraction (EBSD) image quilty (IQ)—inverse pole figure (IPF) maps of the as-hot-rolled specimen showing the boundaries of prior austenite grains (white lines); **b** 3D-APT reconstructured specimen containing two grain boundaries; the needle-shaped specimen was milled at the region marked by the yellow box in **a**; **c** Elemental profiles obtained across the grain boundaries [13]

uncertainty of nuclei density in the nucleation stage. Three steel grades were investigated, namely 0.023C-0.17Mn, Fe-0.1C-0.49Mn, Fe-0.1C-1.0Mn. The ferrite–austenite interface is retarded by the local Mn enrichment and Mn-spike is predicted to stay behind the interface as shown in Fig. 6.4 [14]. The Mn profiles were simulated using the well-known DICTRA software with either LE or PE conditions. In LE model, the interface is assumed to migrate under full LE with the partitioning of both C and alloy element M. The transformation rate in LE model is determined by either Mn or C partitioning, which depends on the temperature and composition [14]. In PE model, it is assumed that there is no partitioning of M and that the chemical potential of C across the interface is constant. Hence, the transformation is only controlled by C diffusion. The curve shows that there are two Mn spikes in each profile: (i) an extremely sharp Mn spike at the interface, which further indicates that the transformation mode is NPLE; (ii) a residual Mn spike left behind the interface [14]. The simulated residual Mn spikes in different alloys are quite different, and the amplitude of the spike increases with increasing Mn concentration. It is concluded that the growth retardation is caused by the moving interface hitting the residual Mn spike in austenite left behind by the previous thermal cycles in the two-phase region [14]. The residual Mn spike itself is caused by the large difference between the interface migration rate and diffusion rate of Mn at the transition temperatures during the cyclic phase transformations [14]. If the growth retardation is indeed caused by the residual Mn spike in austenite, it can easily be expected that the magnitude of growth retardation would also be affected by Mn concentration in the alloys [14].

Most recently, the migration of ferrite/austenite interface was experimentally observed in Fe-0.0848C-0.47Mn (wt.%) by hot stage TEM technique [17]. The ferrite to austenite transformation is characterized to investigate the interface migration.

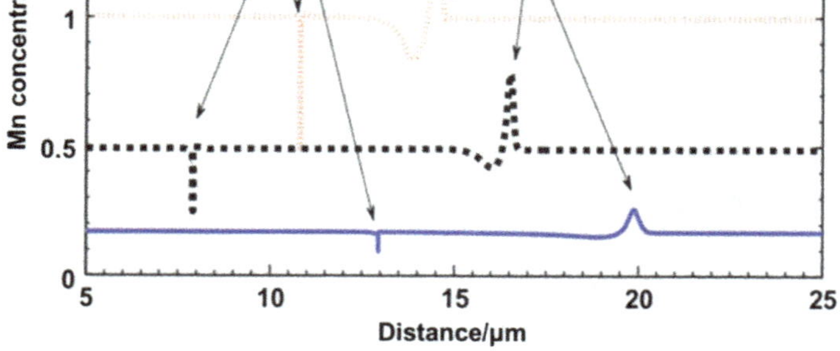

Fig. 6.4 Simulated residual Mn profiles in Fe-Mn-C alloys with different Mn contents after cyclic partial transformation [14]

The annealing temperature ranged around 810 °C for 800 s. The in-situ inspection of moving interface includes the ferrite–austenite stage and inverse stage of Fe-0.1C-0.5Mn steel, shown in Fig. 6.5 [17]. It is not uncommon to observe changing grain boundary structure ahead of, or behind, the ferrite/austenite interface during thermal cycling [17]. In cases where a migrating triple point is observed, competition is seen between the two grains of the growing phase, resulting in the lateral displacement of the triple point and the rapid movement of incoherent grain boundaries to accommodate this displacement [17]. At the end of the transformation, all of the four remaining grains (two austenite and two ferrite) form a single junction, which is only maintained for a short period of time [17].

A recent research by Kuzmina et al. [18] observed that solute Mn directly embrittles martensite grain boundaries while reversion of martensite to austenite at high-angle grain boundaries cleans the interfaces from solute Mn by partitioning the Mn into the newly formed austenite, hence restoring impact toughness, as shown in Fig. 6.6 [18]. Microalloying with B improves the impact toughness in the quenched state and delays temper embrittlement at 450 °C. Tempering at 600 °C for 1 min significantly improves the impact toughness and further tempering at lower temperature does not cause the embrittlement to return [18].

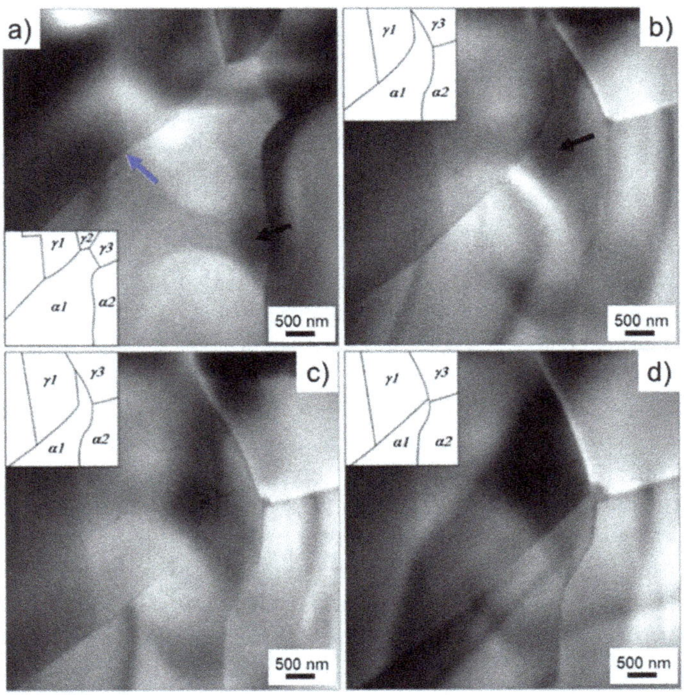

Fig. 6.5 Bright field images of the interface migration obtained from hot-stage TEM during the ferrite to austenite transformation at different states: **a** 708.3 s (811 °C); **b** 740.1 s (816 °C); **c** 751.9 s (818 °C); **d** 756.9 s (819 °C) [17]

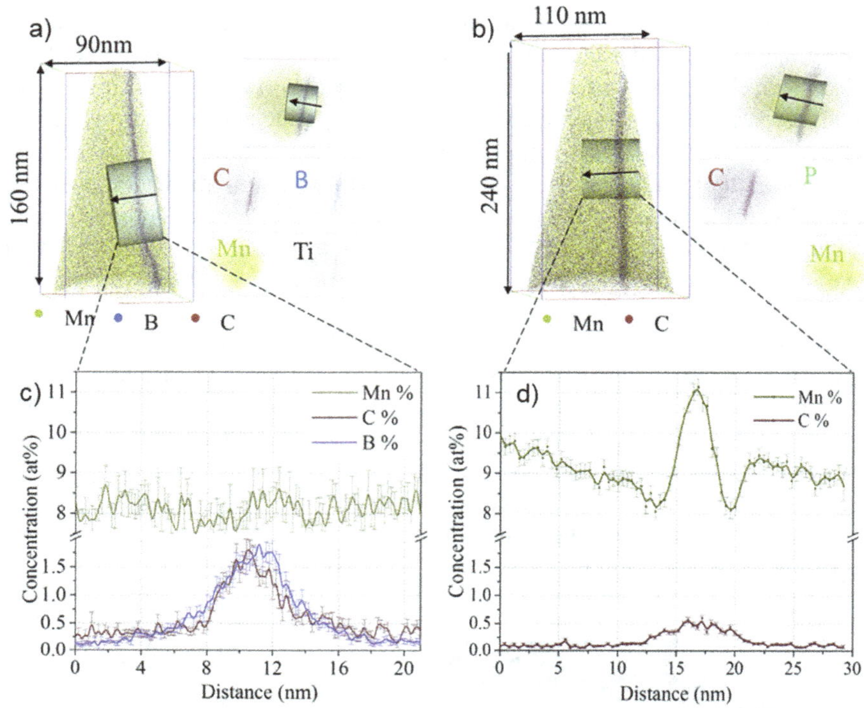

Fig. 6.6 **a** APT results of Mn9 + B steel in the as-quenched state with the corresponding element distribution maps; **b** APT results of Mn9 steel tempered at 450 °C for 1 min with the corresponding element distribution maps; **c**, **d** 1D-elemental profiles obtained by ROI method across the boundaries in **a**, **b** respectively [18]

At higher temperatures, regular austenite nucleation and growth takes place, whereas at lower temperature, Mn directly promotes its growth. The model material with a nominal composition of Fe-0.085C-8.8Mn (wt.%) was chosen to ensure the complete formation of lath martensite upon quenching. The final heat treatment is conducted at 450 °C for 1 h [18]. The microalloying element B facilitates the relieving of harmful segregation which is clearly illustrated in Fig. 6.6. A delay in embrittlement in alloy Mn9 + B is due to the non-equilibrium segregation of B to prior austenite grain boundary (PAGB) (Fig. 6.6a, c), which decreases the grain boundary energy and reduces the diffusion of Mn to PAGBs [18]. B desegregation during tempering in conjunction with promoted segregation of Mn causes an embrittling effect after longer holding times. An increase in impact toughness after longer tempering times is explained in terms of austenite reversion and the associated removal of solute Mn from the grain boundaries into the newly formed austenite [18]. A double-step heat treatment (600 °C plus 450 °C) significantly improved the impact toughness by accelerating the austenite reversion kinetics at the higher temperature and changing the Mn segregation from grain boundaries into the austenite at the lower temperature [18]. The underlying embrittling mechanism of Mn is attributed to, namely, a decrease in grain boundary cohesion by solute Mn (Fig. 6.6b, d) and the influence of local magnetic contributions to grain boundary cohesion, causing a local volume change and internal stresses [18].

Despite the side effect of segregation embrittling materials, segregation could also be engineered to improve the mechanical performance. The term segregation engineering (SE), which is schematically shown in Fig. 6.7a, describes a self-organized microstructure manipulation method that is based on the combination of (i) confined thermodynamically driven chemical Gibbs segregation via solute decoration to specific lattice defect traps, and (ii) the associated structural, order or phase changes triggered at these decorated sites [19]. The concept of SE works particularly well for steels and Ti alloys owing to their rich spectrum of available allotropic phase transformations [19]. Following is the case where local solute segregation, and the associated ordering phenomena at lattice defects, lead to confined martensite-to-austenite reversion in a Fe-9Mn-1.9Ni-0.6Mo-1.1Ti-0.33Al-0.1Si-0.05 steel, as shown in Fig. 6.7b, c. The newly termed approach of jointly manipulating the structure and composition of grain boundaries set great importance in interface engineering [19]. In tempered specimens (450 °C, 48 h), layers with a thickness of 10–15 nm are observed at the martensite block boundaries and at the martensite lath boundaries. These zones are mostly devoid of precipitates, which as a rule only form inside martensite grains (see Fig. 6.7b) [19]. Exceptions can apply in cases where the particle first formed inside the martensite before the zone surrounding it became transformed to austenite. The inset in Fig. 6.7b shows an enlarged image of such a region. The micrograph reveals that this interface zone is characterized by a nano-scaled double layer, where the two individual films are separated from each other by an internal interface. This interface is not visible over the entire length of the double layer in one single micrograph [19]. It appears at different positions along the grain boundary when tilting the sample. This effect is due to a slight bending of the TEM foil, which causes minor orientation differences along the longitudinal axis of the interface with respect to the incident electron beam. Figure 6.7c shows a schematic illustration of the double layer interface [19].

The nano-twin (NT) is strongly thought to be a desirable structure for overcoming the dilemma between strength and ductility. The strength-ductility synergy of NTed materials is controlled by the deformation mechanisms, involving dislocation pile-up at the twin boundary (TB), partial dislocation slips along the TB, and TB migration, or detwinning [20]. The most cutting-edge research by Chen et al. constructed a clear deformation pattern of NTs in type 304 SS austenitic stainless steel by combining in-situ tensile tests with a dislocation-based theoretical model and molecular dynamics simulations, which aided in the design of advanced steels with superior strength-ductility synergy [20]. As an example, the intersected NTs are selected to demonstrate the typical deformation behaviours of successive twinning and detwinning, as shown in Fig. 6.8. The samples with NTs are produced by surface mechanical attrition treatment (SMAT). The intersected bundles of NTs (Fig. 6.8a) have two systems of TB orientations, where the TB1 exhibits a 70° angle to the loading direction and TB2 is approximately parallel to the loading direction [20]. In Fig. 6.8, the red solid lines and arrows indicate the twinning positions and the blue dotted lines and arrows

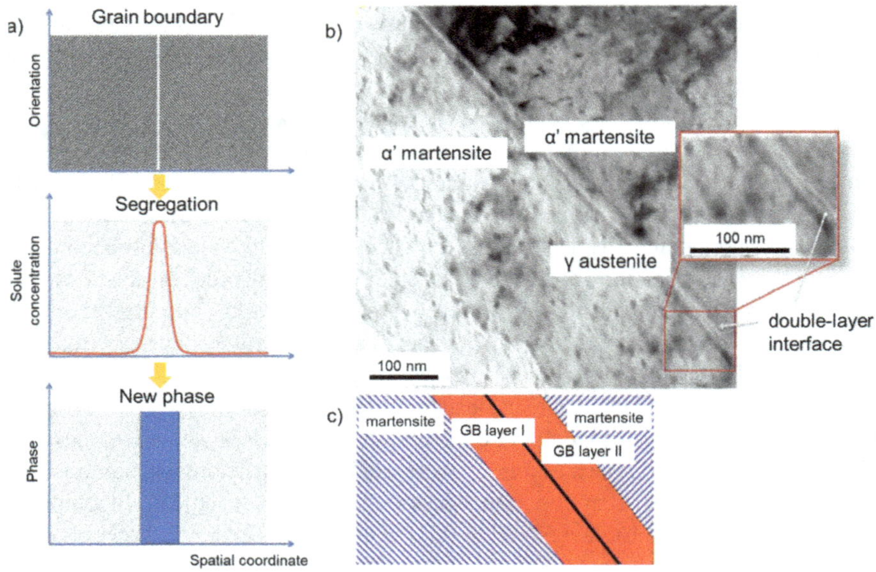

Fig. 6.7 **a** Principle of "segregation engineering", the approach is based on grain boundary segregation followed by nano-scale phase transformation of the decorated grain boundary region; **b** TEM bright field image of the interfacial double layer observed at a martensite-martensite grain boundary; **c** Schematic image of the observed grain boundary layers [19]

illustrate the detwinning positions. For instance, NTs 1–3 along the TB2 direction and NT 4 along the TB1 direction exhibit partial twinning, as shown in Fig. 6.8a, b [20]. Similarly, detwinning also occurs, as shown by NTs 2 and 3 (Fig. 6.8b, c) along the TB2 direction, and NTs 5 (Fig. 6.8b, c), 6 and 7 (Fig. 6.8c, d) along the TB1 direction. In addition to these marked NTs, other twinning and detwinning behaviours are indicated in Fig. 6.8c, d [20]. The statistical results show that co-activation of twinning and de-twinning occurs in the intersected NT bundles with $\lambda < 5$ nm under shear stress, and the NTs with $\lambda = 2$–3 nm contribute to the highest ratio of 45% [20].

Fabricating nano-twin can be thus considered as an important methodology of interface engineering. The statistical results show that 80% of NTs with $\lambda < 5$ nm (five observed zones) exhibit coactivation of twinning and detwinning under shear stress, including intersected and primary NTs [20]. 100% of NTs with $\lambda < 5$ nm (two observed zones) exhibit detwinning and the subsequent martensitic transformation when the TBs are closely parallel to the tensile direction. Secondary twinning occurred in primary NTs ($\lambda = 6$–129 nm) with a frequency of 80% among the five observed zones. Only dislocation motion is observed in the left two zones, and the λ of these twins are at the submicrometre scale [20]. When λ is smaller than 5 nm, the twinning and detwinning are jointly activated under shear stress and detwinning and martensitic transformation occur successively when the TBs are closely parallel to the loading direction [20]. When λ is 6–129 nm, secondary twinning is induced

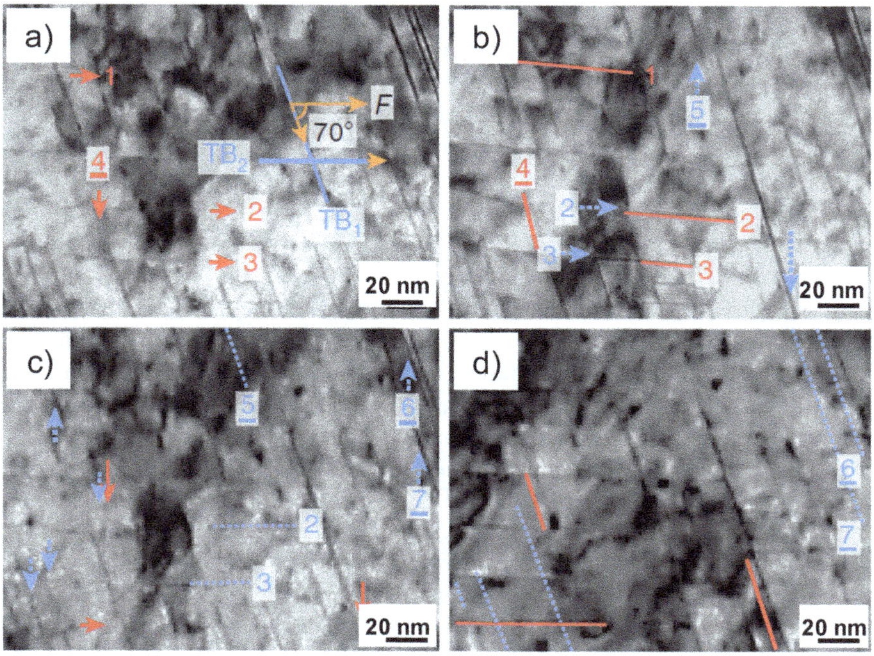

Fig. 6.8 Bright field TEM images of the intersected NTs, showing the deformation evolution during the in-situ tensile tests at different time frames: **a** 0 s; **b** 12 s; **c** 27 s; and **d** 52 s [20]

inside the primary twin system, and the twinning and detwinning are coactivated only in the secondary twin system [20]. As λ increases to the submicrometre scale, dislocation motion is the featured deformation behaviour, as observed in TEM tensile tests, where the dislocations emit from the TBs and propagate normal to the loading direction in the form of dislocation walls. The deformation map of multi-scale NTs provides a clear strengthening mechanism for microstructure manipulation [20]. In contrast to the conventional dislocation strengthening in coarse grains, the twinning, including the secondary twinning, can simultaneously enhance the strength by dislocation pile-up at the TBs and improve the ductility due to partials slipping across TBs [20]. The detwinning softens the material by decreasing the TBs, which is similar to the effect of grain growth [20]. Thus, the coactivation of twinning and detwinning can sustain successive hardening during plastic deformation. This is a special deformation behaviour of NTs on the very fine scale. Accompanied by the detwinning, martensitic transformation is initiated earlier than the other coarse austenite grains, which can strengthen the materials [20].

The correlative study of TEM and APT provides the accuracy in relating microstructural information with crystallographic analysis. In particular, the use of scanning nano-beam diffraction provides valuable complementary crystallographic information when being performed on atom probe tips [21]. This technique enables the measurement of orientation and phase maps as known from electron backscattering diffraction with a spatial resolution down to one nanometre [21]. Herbig et al. investigated a binary Fe-9Mn (wt.%) martensitic alloy, annealed for two days at 450 °C to create small austenitic, Mn-enriched precipitates; and a hypereutectoid pearlitic steel with composition Fe-0.98C-0.31Mn-0.20Si-0.20Cr (wt.%), cold-drawn by $\varepsilon = 6.02$ true strain and annealed for 2 min at 400 °C to create a microstructure composed of columnar grains with an average diameter of about 30 nm and some intergranular cementite particles. Figure 6.9 shows an overlay of the carbon distribution as measured by APT with a bright field (BF)-scanning transmission electron microscopy (STEM) image [21]. The parameters for the reconstruction of the APT map are optimized using the STEM image as a template. That there is a difference between true specimen outline and the APT reconstruction parameters that yield the optimum match also implies that sample outlines as measured in the SEM can only serve as a rough guideline for reconstruction and that TEM micrographs which provide additional information of the sample interior are better suited for the adjustment of the reconstruction parameters [21]. The combined information of local grain orientations as measured by nano-beam electron diffraction (NBD), grain boundary planes as measured from the STEM micrographs and the local chemistry measured by APT, enables the investigation of grain boundary segregation and its dependence on the five macroscopic degrees of freedom of the interface, for materials with grain sizes as low as 30 nm [21]. The STEM micrograph is used to optimize APT reconstruction parameters, yielding excellent agreement between STEM and APT [21]. The 12 at.% iso-concentration surface indicated in green denotes the position of the cementite particles. Individual dislocation lines are visible at low-angle grain boundaries. The method thus provides an efficient way to measure a large number of boundaries [21].

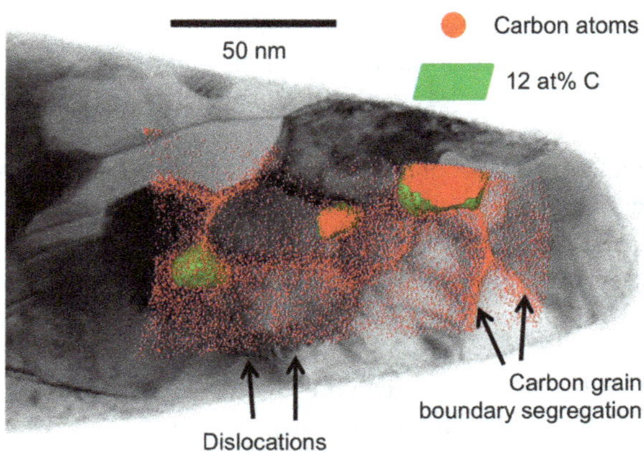

Fig. 6.9 Overlay of BF-STEM micrograph and projected 3D carbon map from APT [21]

6.4 Application

6.4.1 C and Mn Partitioning Across Bcc/Fcc Interface in MMnS Steels

Austenite reverse transformation in medium-Mn steels (MMnS) is associated with the elemental partitioning across bcc/fcc interface during the intercritical annealing which is of great importance for tuning the stability of retained austenite. In order to understand the partitioning behaviour the elements, atom probe tomography was applied to characterize the elemental distribution and partitioning features across the bcc/fcc interface. Figure 6.10 shows the elemental partitioning behaviours of carbon, manganese, and aluminium between ferrite and austenite in the steels with three alloying conditions, i.e. Fe-0.2C-1.2Al-6Mn steel, Fe-0.2C-1.2Al-6Mn-0.05Nb steel, Fe-0.2C-1.2Al-6Mn-0.05Nb-0.22Mo steel. The three steels are termed shortly as 6Mn steel, 6MnNb steel, 6MnNbMo steel.

The 3D atomic map and the distribution of the elements, i.e. Mn, C and Al, at the selected region of interest (ROI) across bcc/fcc interface in the 6Mn steel is shown in Fig. 6.10a. The austenite phase is clearly enriched with Mn and C. In particular, a Mn spike is observed at the bcc/fcc interface and the concentration of the Mn spike is as high as 20 at.%. The average Mn concentration in austenite is about 11 at.% and about 2 at.% in ferrite. The Mn Spike width is about 1–2 nm. In order to examine the effect of single addition of Nb microalloying on the partitioning behaviour of elements, 6MnNb steel was characterized. The 3D atomic map and the distribution of the elements at the selected ROI across bcc/fcc interface in the Nb-microalloyed 6MnNb steel is shown in Fig. 6.10b. A Mn spike at the bcc/fcc interface of 6MnNb is observed as well. The concentration of the Mn spike is about 15 at.%, relatively lower comparing to the one in 6Mn steel and the average Mn concentration in ferrite is about 2 at.%. Furthermore, nano-sized Nb carbides are formed in the ferritic matrix. Nb atoms in the dataset are mainly observed to be enriched in Nb carbides.

The 3D atomic map and the distribution of the elements at the selected ROI across bcc/fcc interface in the multiple-microalloyed 6MnNbMo steel is shown in Fig. 6.10c. Mn spike is also observed in 6MnNbMo steel, however, the Mn concentration (about 12 at.%) is lower than the 6Mn and the 6MnNb steels. The average Mn concentration in ferrite is about 3 at.%, slightly higher than in the other two steels. More specifically, the partitioning behaviour of carbon and aluminum is stronger in the 6MnNbMo steel, comparing with the other two steels. Similarly, nano-sized carbides are observed, and mainly alloyed with Nb and Mo. Decreasing Mn content at the spike with the addition of microalloying elements, i.e. Nb, Mo, indicates that the solute atoms of microalloying elements may hinder the partitioning between austenite and ferrite during phase transformation occurred through intercritical annealing.

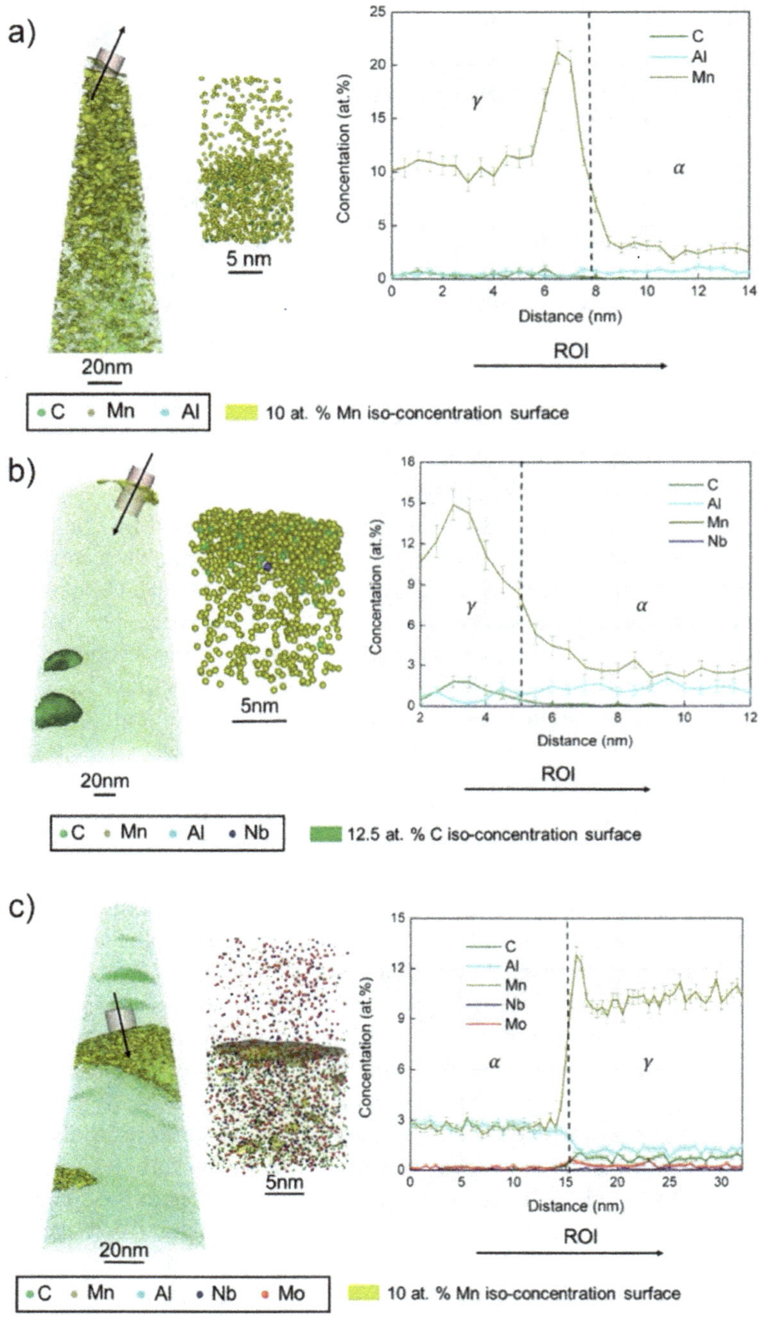

Fig. 6.10 **a** Distribution of the elements at the selected ROI from the phase boundary in the 6Mn steel; **b** Distribution of the elements at the selected ROI from the phase boundary in the sample 6MnNb steel; **c** Distribution of the elements at the selected ROI from the phase boundary in 6MnNbMo steel

6.4.2 *Influence of Microstructural Morphology on Hydrogen Embrittlement*

Hydrogen embrittlement (HE) resistance is an important criterion for modern high-strength steels used in the automotive, power, aerospace and chemical industries. The ultra-fine grain (UFG) duplex microstructure of medium Mn steel consists of a significant amount of austenite and ferrite/martensite, achieving an exceptional balance between mechanical properties and alloying cost. However, the ferrite/martensite phase is considered to be susceptible to hydrogen penetration, which contributes to the severe HE level [22–26]. It has been reported that the austenite, as a critical phase in MMnS to improve mechanical performance, effectively traps the hydrogen atoms and leaves the absorbed hydrogen in a more activated state, leading to brittle fracture by the localized transformation-induced plasticity (TRIP) effect [22, 23, 25]. Han et al. [24] studied the HE behavior in cold-rolled and hot-rolled Fe-7Mn-0.1C steel and correlated the contributions of established HE mechanisms, such as hydrogen-enhanced decohesion (HEDE) [27–29] and hydrogen-enhanced localized plasticity (HELP) [27, 28] with microstructural features. This subchapter aims to study the influence of microstructural morphology on hydrogen-related behaviour, including hydrogen uptake and hydrogen embrittlement [30].

Two heat treatment routes were performed on a cold-rolled medium-Mn steel Fe-12Mn-3Al-0.05C (wt.%), with comparable mechanical properties being obtained. One heat treatment was austenitic reverse transformation annealing (ART) only and the other consisted of a successive combination of austenitizing (AUS) and ART annealing [30]. Figure 6.11 shows UFG microstructure of ART sample characterized by EBSD measurement [30]. Figure 6.11b shows that the austenite (blue area) had two main locations, between the martensite ridges (green area) and in the nodular grain region. The red arrows mark the area with fine-grained spherical structure, while the white arrows indicate the lamellar structure [30]. The grain orientation scattering (GOS) map (Fig. 6.11d) shows that the spherical austenite grains, as identified by the phase map, have low GOS values below 1, which is common for recrystallized grains [31].

Figure 6.12a shows UFG microstructure of AUS + ART sample characterized by EBSD measurement [30]. Figure 6.12b shows the initial microstructure of the AUS + ART sample, which consists of an ultrafine-grained lamellar microstructure and a spherical morphology. After the austenitizing step, the deformed structure is completely eliminated and replaced by quenched martensite. The size of the martensite packages ranges from 1 to 5 μm, the size of the spherical grains decreases to 1 μm after the subsequent ART annealing due to the competition between austenite inversion and martensite recrystallization [30]. As shown in the GOS figure (Fig. 6.12d), the ultrafine-grained martensite colonies exhibit a small scatter in GOS, suggesting that they are recrystallized during the intercritical annealing [30].

The synchrotron X-ray diffraction technique (SYXRD) allows quantitative analysis of the fraction of austenite. Interrupted tensile tests were employed with a termination strain of 5, 15 and 25%. The SYXRD measurements were taken on the uniform

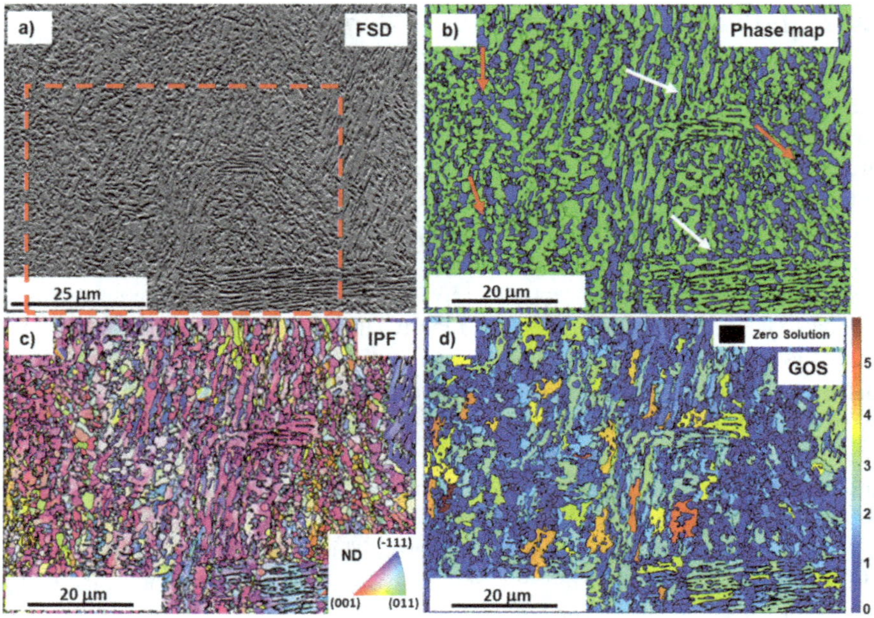

Fig. 6.11 Electron backscatter diffraction measurement of initial microstructure in austenite-reverted-transformation annealed specimen, including **a** Forescatter diode image; **b** Phase distribution map; **c** Inverse pole figure and **d** Grain orientation spread image; The dashed box in **a** denotes the analysed region for **b–d**; red arrows in **b** indicates globular grains; white arrows marks the band-shaped lamellar grains [30]

strain zone adjacent to areas of necking. Figure 6.13 shows the SYXRD profile of samples with increasing degrees of deformation [30]. In the initial microstructure, the ART sample and the AUS + ART sample show a comparable amount of austenite, with a slight difference of ~2%. Due to the relatively low mechanical stability, the austenite transforms into martensite during deformation, which leads to strain hardening behaviour. At the end of the deformation (engineering strain from 25–30.1%), the amount of transformation of austenite to AUS + ART is still high, falling from 40.4–36.4%. Conversely, extremely low austenite transformation of 40.4–40.2% is observed in ART samples with an engineering strain of 25% to 33.1%, indicating that the TRIP effect hardly contributed to the deformation at this stage [30].

Figure 6.14a shows the engineering stress-engineering strain curves of the ART and AUS + ART samples [30]. The mechanical properties of ART and AUS + ART specimens are also comparable. The ART sample shows a slightly higher total elongation of 33.1% and a relatively lower ultimate tensile strength (UTS) of 811 MPa. The yield point is 680 MPa. The AUS + ART sample has an improved yield strength of 701 MPa and an improved UTS of 891 MPa with a total elongation of just over 30% [30]. Figure 6.14b shows true stress–strain curves and the corresponding strain hardening curves of medium Mn steel Fe-12Mn-3Al-0.05C after ART and AUS + ART heat treatment [30]. Three stages can be clearly distinguished in the

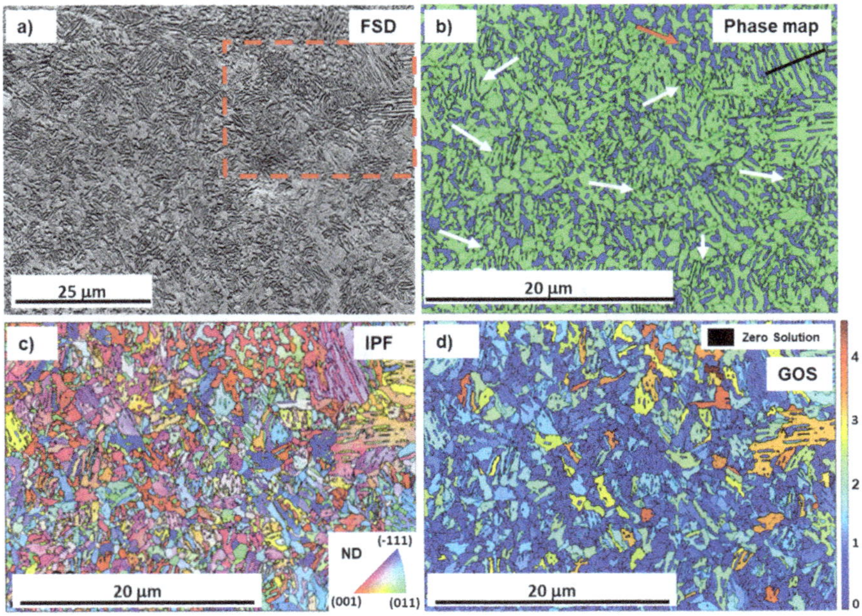

Fig. 6.12 Electron backscatter diffraction measurement of initial microstructure in AUS + ART annealed specimen, including **a** Forescatter diode image; **b** Phase distribution map; **c** Inverse pole figure; and **d** Grain orientation spread image. The dashed box in **a** denotes the analysed region for **b–d**; red arrows in **b** indicates globular grains; white arrows marks the band-shaped lamellar grains [30]

curves based on their distinct characteristics. In stage I, the strain hardening rate falls from an extremely high value. The ART sample finishes Stage I earlier than the AUS + ART sample. In Stage II, strain hardening rates increase during plastic deformation before Stage III strain hardening rates decrease up to the UTS point. In Stage III, final values of strain hardening rate are almost the same for two samples, around 1000 MPa [30].

To assess susceptibility to hydrogen embrittlement, ART specimens and AUS + ART specimens with different H-charging states were subjected to *ex-situ* slow strain rate tensile (SSRT) testing. The strain rate was at 1×10^{-6} s^{-1} to ensure that the fracture behaviour was predominated by the delayed hydrogen-induced cracks [32]. Figure 6.15 shows the mechanical properties before and after the hydrogen ingression. The HE susceptibility can be evaluated by the deterioration of the mechanical properties. The hydrogen content shows little influence on the strength of both sample groups. The yield strengths remain unchanged while the UTS points are not reached before hydrogen-induced brittle fracture [30]. Moreover, the different responses of ductility to hydrogen ingression are observed. The ART sample experiences a clear transition from ductile to brittle with increasing hydrogen uptake. The elongation losses increase with increasing hydrogen charging time in ART sample [30]. In contrast, the presence of even a small amount of hydrogen in the AUS

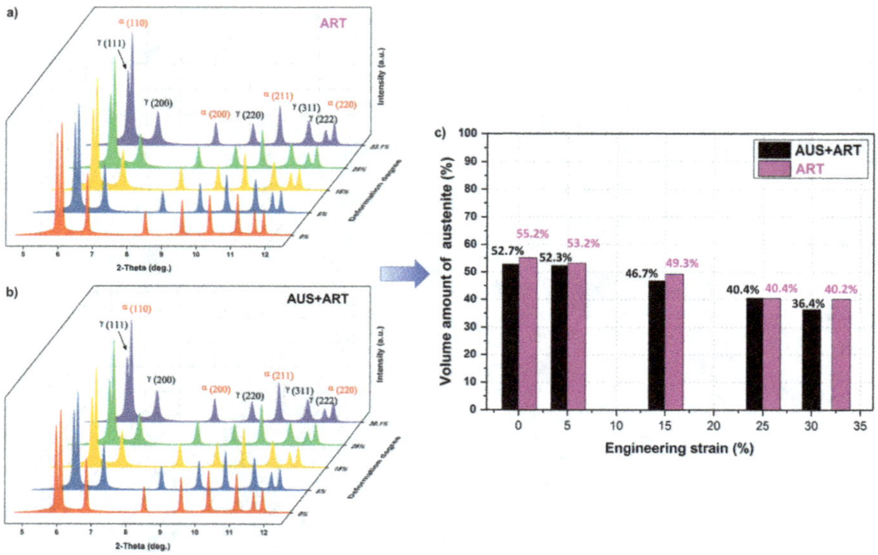

Fig. 6.13 Synchrotron X-ray diffraction profile of **a** Austenite-reverted-transformation specimen; **b** AUS + ART specimen with increasing deformation degree; **c** Volume fraction of austenite in samples in the as-annealed condition as well as taken from interrupted tensile tests at different strains [30]

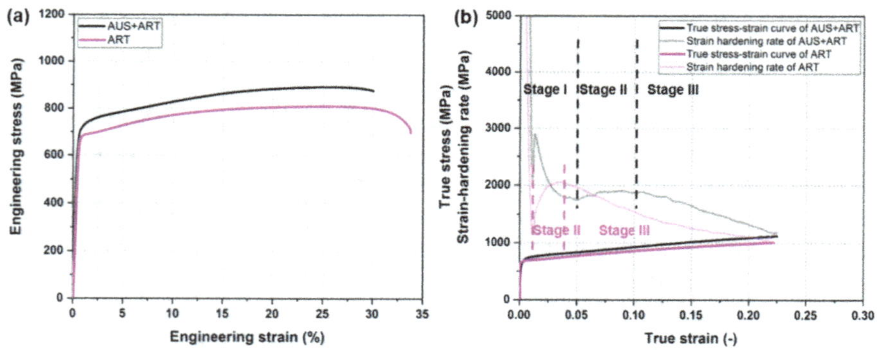

Fig. 6.14 **a** Engineering stress-engineering strain curves; **b** True stress-true strain curves and strain-hardening curves of the medium-Mn steel Fe-12Mn-3Al-0.05C after heat treatment austenite-reverted-transformation (ART) and AUS + ART [30]

\+ ART sample will result in catastrophic failure. When charged with hydrogen, the sample fractures very early with a limited total elongation of ~4%, independent of the hydrogen content [30]. The prediction of the hydrogen embrittlement mechanisms focuses on two main aspects, hydrogen absorption and fracture mode [30].

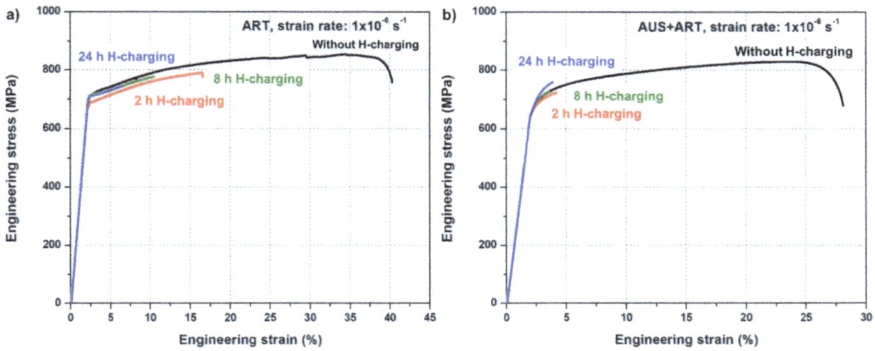

Fig. 6.15 Evaluation of mechanical degradation by ex-situ slow strain rate tensile test at a strain rate of 10^{-6} s^{-1} in **a** ART specimen; and **b** AUS + ART specimen [30]

Considering the similar austenite-ferrite ratio in the initial microstructure, the divergence in hydrogen absorption could be attributed to the different microstructure morphologies [30]. The premature fracture in all AUS + ART specimens with different hydrogen contents manifests that the amount of hydrogen is not the critical control factor. By correlating the hydrogen desorption curve with the microstructure (Fig. 6.16), the increased hydrogen content and catastrophic failure can be correlated with the large interface density in AUS + ART [30].

The microstructure of AUS + ART consists mainly of fine martensite colonies with a package size of 3 μm, offering a variety of interfaces. These interfaces encompass the martensite lath boundaries, prior austenite grain boundaries and phase boundaries. A large quantity of interfaces offers more trapping sites for hydrogen atoms [30]. Meanwhile, the interface may act as the fast diffusion path for hydrogen transport. The increased hydrogen concentration in AUS + ART samples could be attributed to the trapping and rapid diffusion behaviour of hydrogen atoms. The

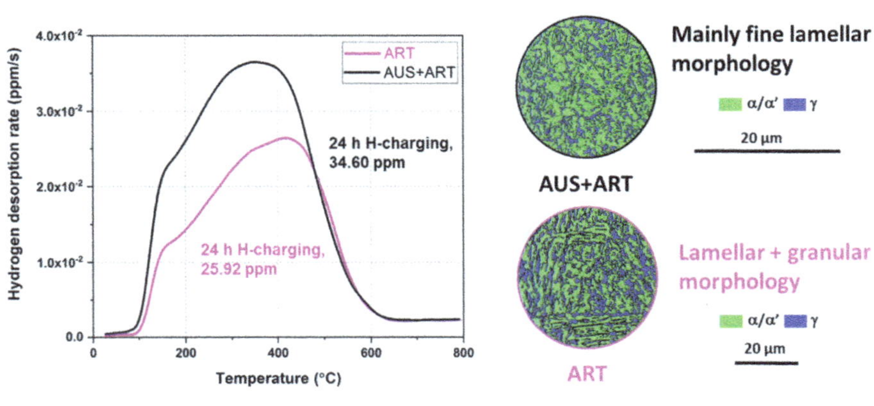

Fig. 6.16 Hydrogen desorption rate as a function of temperature and the corresponding microstructural features [30]

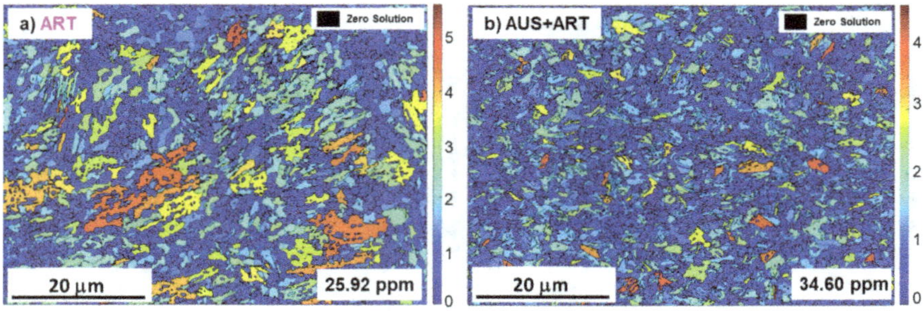

Fig. 6.17 Grain orientation spread figures of **a** Austenite-reverted-transformation (ART) and **b** AUS + ART specimens undergone slow strain rate tensile test after 24 h hydrogen charging [30]

different fracture properties in ART and AUS + ART samples result from the contributions of different hydrogen embrittlement mechanisms [30]. In ART samples, film-like austenite grains with high aspect ratio can still transform (warm-coloured area in Fig. 6.17a) even after strong hydrogen input, which is probably related to the HELP mechanism. The austenite film shows a high GOS within the grains. Therefore, jagged facets can be observed due to recurrence of localized TRIP effect and HEDE mechanism [24]. The deformation-induced martensitic transformation is considered to be prone to HE because of the localized deformation and crack initiation [22, 23, 25]. In the strongly hydrogen-affected sample (hydrogen loading for 24 h), the HEDE mechanism possibly becomes the dominant effect, reducing the cohesive bond strength of the lattice [24, 29]. In addition, it was reported by Han et al. [24] that crack propagation can be impeded in the granular region, since cracks often have to change the propagation direction when entering the nearby phase.

It is generally believed that the hydrogen trapped at interfaces decreases the bond strength according to the HEDE mechanism. A large number of interface defects in the UFG microstructure directly provides more chances for crack initiation [30]. The reduced bond strength can also facilitate dislocation nucleation and emission according to the absorption-induced dislocation emission (AIDE) mechanism [27, 33], leaving small and shallow dimples at the fracture surfaces. Localized microstrain (in the red region in Fig. 6.17b) drives dislocations to interact with voids and initiate the cracks. The large interface provides a significant opportunity for cracking in AUS + ART samples, resulting in poor resistance to hydrogen embrittlement [30].

In addition to the experimental observations, qualitatively conceivable relevant failure scenarios are considered. While kinetic properties, especially interfacial diffusion, are expected to play an important role, the focus here is on thermodynamic aspects. The grain boundary (GB) distribution in the different microstructures, which differs depending on the type of heat treatment, is decisive. To relate the variation in GB distribution to divergences in the formation of hydrogen-rich phases, the local elastic modulus of the GBs is connected to the modification of the local chemical equilibrium at the GBs [30]. As shown in related work in [34] where the modification of hydrogen solubility in iron due to elastic effects, the shifted solubility limit of hydrogen can be described as a result of modified elastic relaxation of the coherence stresses near and at interfaces the precipitation of the hydrogen-rich phase will result in localized nano-hydride formation [35] While precipitates near a free surface respond less elastically to coherence stresses, a rigid surface presents the opposite case, showing reduced elastic relaxation of coherence stresses near or at the interface. The reduction in hydrogen solubility limit can be as much as two orders of magnitude compared to room temperature bulk sites [30].

To connect these fundamental findings with the experimental observations, the role of the local elastic properties of GBs in iron should be highlighted [36]. Although limited to symmetric GBs, the presented local Young's modulus data of different grain boundaries are used as representatives of high-angle and low-angle grain boundaries. It should be noted that the range of local Youngs modulus in the GB plane is 130–176 MPa in the GB, with the minimum associated with the lowest tilt angle (about 39°), the maximum with the highest tilt angle (approx. 109°) [30]. It can be thus expected that a higher proportion of low-angle GBs, coupled with a lower local Young's modulus, will result in significantly reduced hydrogen solubility and therefore promote hydriding and failure. Having established this connection between GB character and local Young's modulus, it is necessary to relate the GB distribution to the hydrogen-induced failure of the samples. For the UFG microstructure, based on our considerations, we expect low angle grain boundaries to support the precipitation of hydrogen-rich phases and not to suppress them. Consequently, the hydrogen failure in the UFG sample in terms of hydrogen ingression can be explained also from a thermodynamic point of view [30]. As the relaxation of the associated coherency stresses increased, a higher fraction of low angle grain boundaries in AUS + ART specimen is responsible for the formation of hydrogen-rich phases. From a thermodynamic point of view, the increased hydrogen concentration in AUS + ART samples is attributed to the rapid ingress of hydrogen and the precipitation of brittle hydrogen-rich phases, leading to premature failure [30].

In this subchapter, the mechanical properties and susceptibility to hydrogen embrittlement on a medium-Mn-Fe-12Mn-3Al-0.05C cold-rolled steel were studied [30]. A combination of AUS and ART produces comparable mechanical properties to those in a routine where the ART anneal was applied merely after cold rolling. The ultrafine-grained martensite colonies provide a large number of interfaces (former austenite boundaries and lath boundaries) for hydrogen trapping, increasing hydrogen penetration. The ART sample shows a clear ductile–brittle transition with increasing hydrogen concentration. Hydrogen embrittlement is considered to be dominated by the simultaneous contribution of the HEDE and HELP mechanisms. The extremely high hydrogen susceptibility of AUS + ART samples is mainly attributed to the HEDE mechanism in the UFG microstructure with a large number of interfaces. Accounting for thermodynamic factors suggests that the error discrepancy in the UFG and non-UFG samples is likely related to the facilitation of hydrogen-rich phase precipitation by interface defects [30].

6.4.3 Carbon Enrichment at Nano-Twin Boundaries in Bearing Steels

100Cr6 bearing steel is extensively studied and used in the manufacturing, transportation and construction industries over the last several decades. Bearing steels are exposed to extreme conditions throughout their service life and suffer from rolling contact fatigue. Consequently, damage builds up below the contact surface and leads to disadvantageous structural changes such as dark etched areas, white etched areas or white etched areas [37]. In this sub-chapter, the effects of quenched martensite in high-C steels on microstructural evolution was reported, with a focus on the role in elemental partitioning and precipitation during tempering. The microstructure of the as-quenched plate martensite in a high-C steel 100Cr6 was characterized by means of electron microscopy and atom probe tomography. Carbon redistribution behaviour was studied at the atomic level, revealing the nature of the transformation dynamics that were affected by carbon and other substitutional alloying elements [37].

The investigated 100Cr6 steel, Fe-0.967C-0.3Si-0.23Mn-1.38Cr (in wt%), was hot forged and annealed at 800 °C for 7200 s. The samples were quenched to form a spheroidized structure, then heated and austenitized at 850 °C for 300 s, followed by quenching to room temperature in an Ar atmosphere. The micrographs are shown in Fig. 6.18 [37]. Figure 6.18a reveals the spheroidized microstructure prior to the austenitization of 100Cr6, which consists of ferritic matrix and spheroidized carbides. After quenching, the microstructure exhibits a martensitic matrix with partially dissolved spheroidized carbides (Fig. 6.18b). The volume fraction of carbides is estimated from the area fraction in the scanning electron microscopic images by the measuring at least 2000 particles [37]. The carbide fraction reveals 14 and 3 vol.% before and after austenitization, respectively. During austenitizing, the average carbide size decreases from 0.63 ± 0.02 μm to 0.49 ± 0.02 μm. After austenitization at 850 °C for 300 s, most of the spheroidized carbides are dissolved and only 3 vol.% remain with an average size of 0.49 ± 0.02 μm. After quenching, the structure shows a martensitic matrix with partially dissolved spherical carbides (Fig. 6.18c, d) [37].

The partially dissolved spheroidized carbides are identified to be $(Fe,Cr)_3C$ from crystallographic analysis by TEM and from chemistry analysis by APT technique. Bright field TEM images in Fig. 6.19 display a martensitic microstructure in the as-quenched state [37]. The plate-like martensitic microstructure is characterized by $\{112\}_\alpha$ nanotwins. Figure 6.19a shows the two adjacent domains of parallel-aligned twins separated by their midrib. The higher magnification TEM image in Fig. 6.19b allows the determination of the twin length and thickness, which are in the range of 100–500 and 5–10 nm, respectively [37].

Figure 6.20a shows the reconstructed atom probe tip of the as-quenched sample [37]. The enrichment zones of carbon can be recognized by 7.0 at.% carbon iso-concentration surfaces in the elemental map. The carbon appears to be concentrated at the GB and at twin boundaries within the martensite plate. In addition, the formation of clusters of dissolved carbon (8–15 at.% carbon) is visible. Carbon enrichment on the GBs gives an average concentration of about 6.5 at.% [37]. Within the martensite

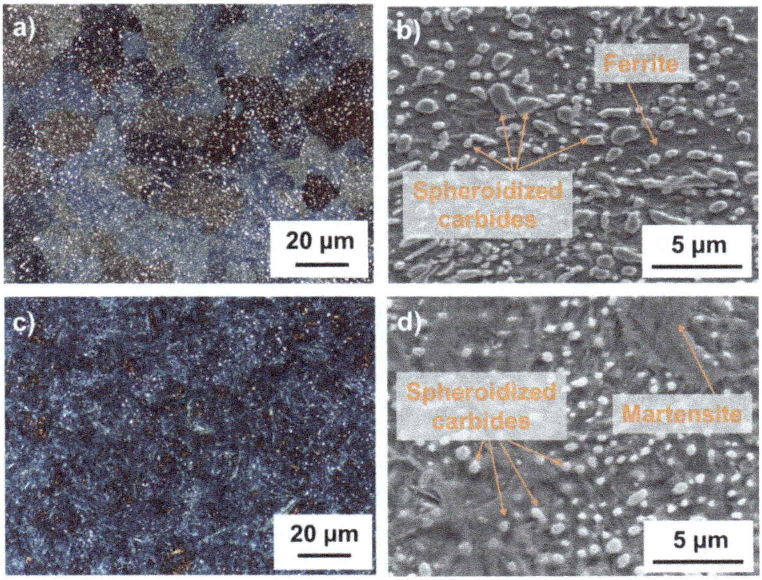

Fig. 6.18 Metallography and microstructure of 100Cr6 bearing steel under different conditions: **a** LOM; ferritic matrix with spheroidized carbides after spheroidization at 800 °C for 7200 s; **b** SEM; ferritic matrix with spheroidized carbides at 800 °C for 7200 s; **c** LOM; martensitic matrix with partially dissolved spheroidized carbides after austenitization at 850 °C for 300 s and the subsequent quenching in Ar; and **d** SEM; martensitic matrix with partially dissolved spheroidized carbides at 850 °C for 300 s and the subsequent quenching in Ar [37]

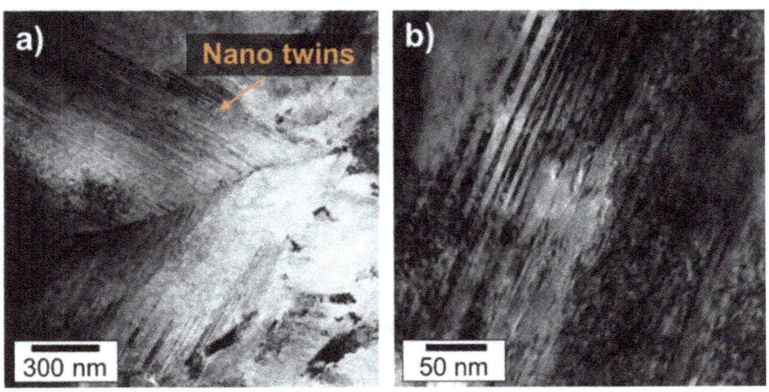

Fig. 6.19 **a** Bright-field micrographs of the intersection of two martensitic twin domains separated by a twin midrib in 100Cr6 steel in the as-quenched state; and **b** Nano-twins in 100Cr6 [37]

plates, the carbon is segregated into layers, indicating the twin pattern in martensite. The atomic concentration profiles along the z-axis of the marked ROI in Fig. 6.20b illustrates periodic carbon spikes, corresponding to C-segregated twin boundaries. The corresponding 1D-concentration profile in Fig. 6.20c facilitates the identification of twin boundaries through the peaks in carbon concentration. The C-content at the twin boundaries ranges from 3.5–5.5 at.% with an approximate average of 4.5 at.%. On this basis, an average twin thickness is measured as 5.7 nm ± 1.2 nm [37]. The substitutional elements such as Si, Mn and Cr show no evidence of segregation at twin interfaces, as indicated by the 1D concentration profile. The calculated 2D concentration map in Fig. 6.20d confirms, in addition to the local C enrichment within the twin boundaries of a maximum of 6 at.%, the carbon segregation to the twin boundaries [37]. The substitutional elements, such as Si, Mn and Cr show no signs of segregation at the twin interfaces, which indicated by the 1D concentration profile in Fig. 6.20b [37].

A detailed examination of the distribution of alloying elements within the ROI of Fig. 6.20 is provided in Fig. 6.21 [37]. A 2D concentration contour map (Fig. 6.21a) is separated into the atomic distribution of interstitial carbon atoms (Fig. 6.21b, c) and the substitutional elements of Si and Mn (Fig. 6.21d, e). As mentioned above, carbon shows a periodic segregation that agrees well with the experimentally determined distance of twin boundaries. As a precaution and to verify previous observations, a 2D concentration distribution of carbon atoms within the ROI is quantified (Fig. 6.21c) [37]. The 2D concentration contour map observes the carbon segregation near the twin interface and a carbon blip at the twin boundary interface. The maximum carbon concentration at twin boundaries locally reaches 6 at.%. Furthermore, the substitutional elements in 100Cr6, such as Si and Mn, reveal a homogeneous distribution along the ROI (Fig. 6.21d, e) [37]. The experimental results indicate that carbon atoms determine the extent of twinning since they promote strong lattice distortions during the martensite formation process [37].

6.5 Summary

As a newly proposed methodology in the last decade, interface engineering in advanced alloys attracts tremendous research interests. The interfaces, on one hand, may act as obstacles to impede dislocation motion, on the other hand, control the transformation behaviour during heat treatment. As the result of influencing the partitioning behaviour and transformation kinetics, the obtained microstructure can be tailored to satisfy service property requirements. Along with many advantages brought by introducing interfaces, the risk of several materials problems, e.g. crack initiation and hydrogen embrittlement, can also be accompanied. Interface engineering approach exhibits huge potential for its application in developing novel metallic materials yet needs to be comprehensively studied in the future.

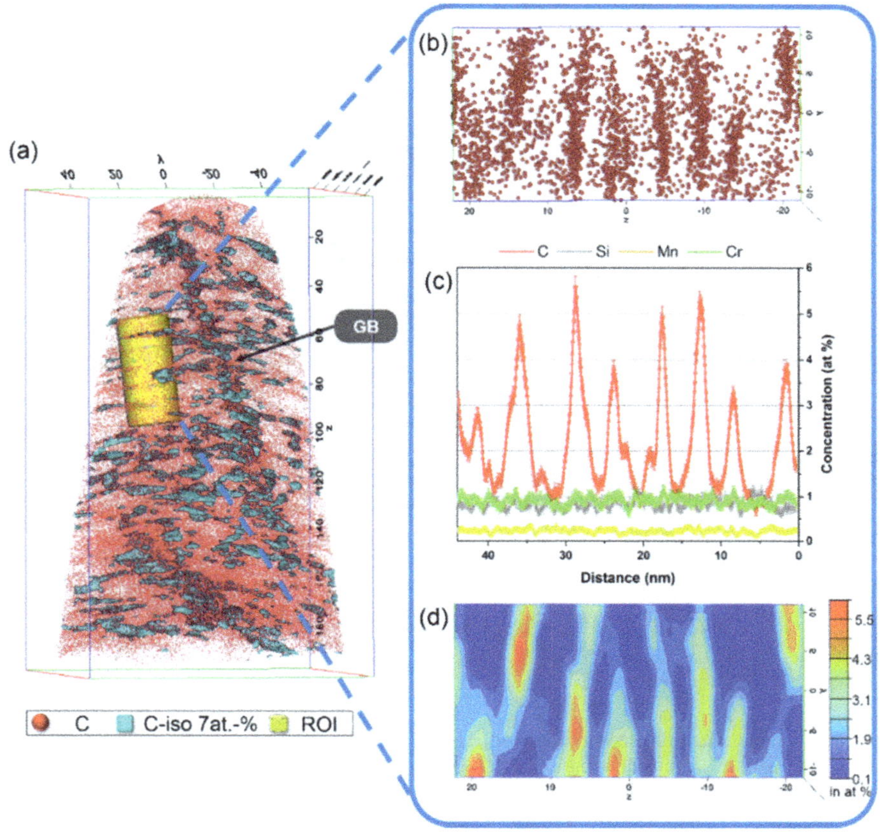

Fig. 6.20 **a** 3D carbon atomic map, 7 at.% iso-concentration surfaces and marked ROI in 100Cr6 in the as-quenched condition subsequent to austenitization; **b** The atomic illustration of carbon along the z-axis of ROI; **c** The corresponding 1D-concentration profile; and **d** The corresponding 2D-concentration map; GB represents grain boundary [37]

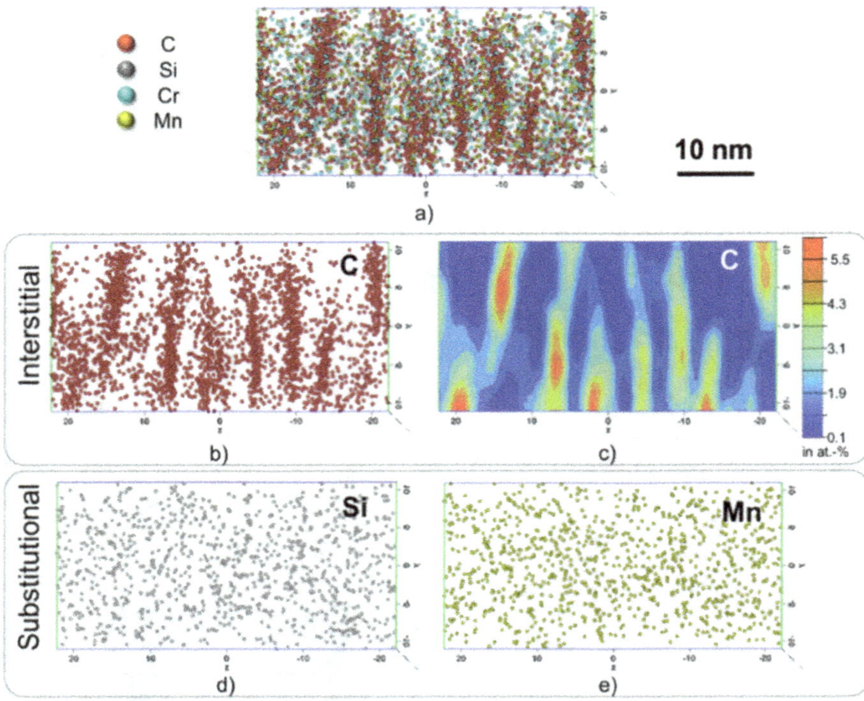

Fig. 6.21 **a** Overview of the alloying elements C, Si, Mn and Cr within the ROI in form of a 2D-atomic map; **b** C atoms distribution; **c** 2D concentration map of C; **d** Si atoms distribution; and **e** Mn atoms distribution [37]

References

1. Mehl R (1983) The historical development of physical metallurgy. In: Cahn R, Haasen P (eds) Physical metallurgy, Amsterdam, North-Holland Physics Pub
2. Aust K, Chalmers B (1970) Structure of grain boundaries. Metall Mater Trans B 1:1095–1104
3. Cahn R (2001) Intermetallics: new physics. Contemp Phys 42:365–375
4. Watanabe T (2011) Grain boundary engineering: historical perspective and future prospects. J Mater Sci 46:4095–4115
5. Davies P, Randle V (2001) Grain boundary engineering and the role of the interfacial plane. Mater Sci Technol 17:615–626
6. Ananya M, Andrea P (2011) Controlling properties of HSLA line pipe steels for oil and gas applications. In: Proceedings of the 6th international conference on high strength low alloy steels (HSLA Steels' 2011)
7. Collins L, Godden M, Boyd J (2014) Microstructures of linepipe steels. Can Metall Q 22:169–179
8. Pattabhiraman H (2013) Kinetics of austenite to ferrite transformation and microstructure modelling in steels. Master thesis, Delft University of Technology
9. Li Z, Ding H, Misra R, Cai Z (2017) Microstructure-mechanical property relationship and austenite stability in medium-Mn TRIP steels: the effect of austenite-reverted transformation and quenching-tempering treatments. Mater Sci Eng A 682:211–219

10. Kamoutsi H, Gioti E, Haidemenopoulos G, Cai Z, Ding H (2015) Kinetics of solute partitioning during intercritical annealing of a medium-Mn steel. Metall Mater Trans A 46A:4841–4846
11. Tirumalasetty G, van Huis M, Fang C, Xu Q, Tichelaar F, Hanlon D, Sietsma J, Zandbergen H (2011) Characterization of NbC and (Nb, Ti)N nanoprecipitates in TRIP assisted multiphase steels. Acta Mater 59:7406–7415
12. Verbeken K, Barbé L, Raabe D (2009) Evaluation of the crystallographic orientation relationships between fcc and bcc phases in TRIP steels. ISIJ Int 49:1601–1609
13. Han J, da Silva A, Ponge D, Raabe D, Lee S, Lee Y, Lee S, Hwang B (2017) The effects of prior austenite grain boundaries and microstructural morphology on the impact toughness of intercritically annealed medium Mn steel. Acta Mater 122:199–206
14. Chen H, van der Zwaag S (2013) Analysis of ferrite growth retardation induced by local Mn enrichment in austenite created by prior interface passages. Acta Mater 61:1338–1349
15. Chen H, Gamsjäger E, Schider S, Khanbareh H, van der Zwaag S (2013) In-situ observation of austenite-ferrite interface migration in a lean Mn steel during cyclic partial phase transformations. Acta Mater 61:2414–2424
16. Chen H, Xu W, Goune M, van der Zwaag S (2012) Application of the stagnant stage concept for monitoring Mn partitioning at the austenite-ferrite interface in the intercritical region for Fe-Mn-C alloys. Philos Mag Lett 92:547–555
17. Nutter J, Farahani H, Rainforth W, van der Zwaag S (2019) Direct TEM observation of α/γ interface migration during cyclic partial phase transformations at intercritical temperatures in an Fe-0.1C-0.5 Mn alloy. Acta Mater 178:68–78
18. Kuzmina M, Ponge D, Raabe D (2015) Grain boundary segregation engineering and austenite reversion turn embrittlement into toughness: example of a 9 wt.% medium Mn steel. Acta Mater 86:182–192
19. Raabe D, Sandlöbes S, Millán J, Ponge D, Assadi H, Herbig M, Choi P-P (2013) Segregation engineering enables nanoscale martensite to austenite phase transformation at grain boundaries: a pathway to ductile martensite. Acta Mater 61:6132–6152
20. Chen A, Zhu L, Sun L, Liu J, Wang H, Wang X, Yang J, Lu J (2019) Scale law of complex deformation transitions of nanotwins in stainless steel. Nat Commun 10:1403
21. Herbig M, Choi P, Raabe D (2015) Combining structural and chemical information at the nanometer scale by correlative transmission electron microscopy and atom probe tomography. Ultramicroscopy 153:32–39
22. Ryu J, Chun Y, Lee C, Bhadeshia H, Suh D (2012) Effect of deformation on hydrogen trapping and effusion in TRIP-assisted steel. Acta Mater 60:4085–4092
23. Wang M, Tasan C, Koyama M, Ponge D, Raabe D (2015) Enhancing hydrogen embrittlement resistance of lath martensite by Introducing nano-films of interlath austenite. Metall Mater Trans A 46:3797–3802
24. Han J, Nam J, Lee Y (2016) The mechanism of hydrogen embrittlement in intercritically annealed medium Mn TRIP steel. Acta Mater 113:1–10
25. Zhang Y, Hui W, Zhao X, Wang C, Cao W, Dong H (2019) Effect of reverted austenite fraction on hydrogen embrittlement of TRIP-aided medium Mn steel (0.1C–0.5Mn). Eng Fail Anal 97:605–616
26. Jeong I, Ryu K, Lee D, Jung Y, Lee K, Lee J, Suh D (2019) Austenite morphology and resistance to hydrogen embrittlement in medium Mn transformation-induced plasticity steel. Scr Mater 169:52–56
27. Lynch S (2012) Hydrogen embrittlement phenomena and mechanisms. Corros Rev 30:105–123
28. Robertson I, Sofronis P, Nagao A, Martin M, Wang S, Gross D, Nygren K (2015) Hydrogen embrittlement understood. Metall Mater Trans A 46:2323–2341
29. Oriani R (1972) A mechanistic theory of hydrogen embrittlement of steels. Berich Bunsen Gesell 76:848–857
30. Shen X, Song W, Sevsek S, Ma Y, Hüter C, Spatschek R, Bleck W (2019) Influence of microstructural morphology on hydrogen embrittlement in a medium-Mn steel Fe-12Mn-3Al-0.05C. Metals 9:929

31. Field D, Bradford L, Nowell M, Lillo T (2007) The role of annealing twins during recrystallization of Cu. Acta Mater 55:4233–4241
32. Henthorne M (2016) The slow strain rate stress corrosion cracking test-A 50 year retrospective. Corros 72:1488–1518
33. Lynch S (1989) Metallographic contributions to understanding mechanisms of environmentally assisted cracking. Metallogra 23:147–171
34. Spatschek R, Gobbi G, Hütter C, Chakrabarty A, Aydin U, Brinckmann S, Neugebauer J (2016) Scale bridging description of coherent phase equilibria in the presence of surfaces and interfaces. Phys Rev B 94:134106
35. Pezold J, Lymperakis L, Neugebauer J (2011) Hydrogen-enhanced local plasticity at dilute bulk H concentrations: the role of H-H interactions and the formation of local hydrides. Acta Mater 59:2969–2980
36. Bhattacharya S, Tanaka S, Shiihara Y, Kohyama M (2014) *Ab initio* perspective of the (110) symmetrical tilt grain boundaries in bcc Fe: application of local energy and local stress. J Mater Sci 49:3980–3995
37. Song W, Drouven C, Galindo-Nava E (2018) Carbon redistribution in martensite in high-c steel: atomic-scale characterization and modelling. Metals 8:577

Chapter 7
Short-Range Ordering Engineering

The aforementioned novel strengthening concepts inspired me to look for new strengthening concepts to improve mechanical properties. The short-range ordering (SRO) concept is therefore recently proposed and applied in various steels as a new pathway to overcome the strength-ductility trade-off in nanostructured metallic materials. This chapter describes the short-range ordering engineering approach and its application in the high strength steels. Own-related research work[1,2,3,4] of precipitation engineering is then reported as application examples in Sect. 7.3.

7.1 Definition and History

The concept of short-range ordering was first proposed by J. Cohen and M. Fine in 1962 [1]. This early article defined the short-range ordering as the local order where the number of short-range unlike atom pairs was greater than that in a random solution [1]. In the alloys without long range order, the probability for an unlike atom pair tends to manifest large interatomic distance, 20–50 Å. With the use of X-ray scattering technology, the short-range order is characterized by superimposed diffraction peaks to analyse the chemistry nature, formation kinetics, formation mechanism and destruction during deformation [1]. The nano-sized ordered structure in metallic materials is specified into two categories: long-range ordering (LRO) and SRO. The LRO refers to the periodical repeating of the spatial and chemical atoms arrangement in a longer distance over nanometre level; while the SRO refers to the structures in which the

Own work-related examples link to the publications:

[1] Song et al. [88].

[2] Drouven et al. [101].

[3] Song and Houston [130].

[4] Song et al. [136].

W. Song, *Nano-Engineering of High Strength Steels*,
Topics in Mining, Metallurgy and Materials Engineering,
https://doi.org/10.1007/978-3-031-42967-5_7

arrangement of atoms is over a distance comparable to interatomic distances. In the newly developed alloy systems, e.g. high-Mn steels and high-entropy alloys, the SRO formation is inevitable, which might generate problems during materials processing (hot cracking during forging/rolling) without an appropriate control. It is of significant importance to understand the SRO formation in nanostructured metals and may further utilize it as an effective strengthening approach.

Using LRO nano-precipitates to gain improved mechanical properties has been studied for over 50 years and it has been widely applied in various metallic systems. The LRO nano-precipitates disperse in the metallic matrix, acting as obstacles to dislocation movement. The strength of the materials can be effectively enhanced, which is also recognized in the materials science and engineering community as precipitation strengthening. The strength increment strongly relies on the volume fraction, the size of nano-precipitates, and the nature of the interaction between the nano-precipitates and dislocations [2]. The strengthening effect of precipitates is usually described by either Kelly-Fine mechanism or Orowan mechanism [2]. According to the Kelly-Fine mechanism, the small precipitates are rather cut by dislocations. In contrast, the large precipitates are circumvented by the Orowan mechanism. The LRO precipitation strengthening is successfully used in various metallic materials, e.g. Al-Cu alloys, Ni-based superalloys etc. In 2015, Kim et al. [3] reported an example of controlling the morphology of LRO B2 phase to improve the strength of steel up to 1.5 GPa. In 2017, another application of the LRO nano-precipitate B2 phase was reported by Jiang et al. [4]. By controlling of the lattice misfit between B2 intermetallic phase and matrix, the strength in maraging steels was largely enhanced. Utilization of LRO nano-precipitates shows great potential in strengthening the metallic materials, yet with a sacrifice on the ductility of the materials.

7.2 Short-Range Ordering (SRO) in Metallic Materials

7.2.1 Formation of SRO in Metallic Materials

SRO in binary and ternary alloy systems have been intensively studied, i.e. Fe-Al alloy [5], Ni-Mo alloy [6], Ti-Al alloy [7], Mg-Zn alloy [8], Al-Cu-Mg alloy [9–12], Al-Mg-Si alloy [13, 14], Ti-Al-V alloy [15] and Ni-Cr-Fe alloy [16]. For example, in ternary Al-Cu-Mg alloys, the rapid hardening reaction during aging was reported to be related to the formation of Cu-Mg SRO in sub-nanometre [16]. The SRO strengthening is believed to be the dominant strengthening mechanism [16–18]. It has been revealed that the nano-scale Cu-Mg SRO are as small as ~0.3 nm [11]. In a Ti-6Al-4V alloy, the modification of the dislocation behaviour by SRO was reported recently [15]. It was found that the SRO can inhibit the cross-slip and promote the formation of pile-ups or pairs of dislocations in this alloy [15]. Beneficial effects of ordering were reported by Neeraj and Mills [7] in a study regarding the occurrence of SRO in an α-Ti-6Al alloy. The SRO formation mechanisms and its response to material processes, however, still remain unclear. Ab initio theoretical calculations have been demonstrated to be an effective approach to investigate the SRO in high-Mn steels in Fe-Mn-C and Fe-Mn-Al-C systems. Kang et al. [17] reported ab initio calculation-based model to describe the yield strength of Fe-Mn-C high-Mn steels considering the influence of SRO on dislocation glide. Sevsek and Bleck [18] further proposed a model to calculate the yield strength of Fe-Mn-Al-C high-Mn steels, including the impact of SRO, grain size refinement, and solid solution strengthening. The results indicate that the SRO contributes to the yield strength as high as approximately 65 MPa in high-Mn steels, which has a significant effect on improvement of yield strength in high-Mn steels.

Fe-Al alloys are of high interest due to their excellent high temperature oxidation-corrosion resistance and whilst having reasonable strength at room temperature [19]. Thermodynamic calculations describe the transitions between the chemically disordered A2 phase through to the B2, B32 and $D0_3$ ordered states, including also ferromagnetic ordering transitions [20]. In the temperature range between 250 and 400 °C and at Al contents between 10 and 20 at.%, there remains inconclusive evidence about the exact nature of the chemical structure [21]. B2 and $D0_3$ domains are formed from the disordered α phase by a continuous α-B2-$D0_3$ ordering transition in the weldment of Fe-Mn-Al-C alloys [22]. The ternary solutes, i.e. Ti, V, Nb, Mo, may also influence the B2-$D0_3$ ordering transition [5].

SRO in Fe-Mn and Fe-Cr alloys upon severe plastic deformation performed by high pressure torsion (HPT) was investigated by Shabashov and Kozlov et al. [23, 24]. The results show that the atomic ordering changes from disordering upon cold (80–298 K) deformation to accelerated short-range ordering upon warm (up to 573 K) deformation when the deformation temperature is changed. The reason for the accelerated short-range ordering is the continuous generation of a large number of mobile point defects. Shabashov et al. [25] investigated the effect of severe plastic deformation and high energy electron irradiation on the binary alloy $Fe_{(100\text{-}X)}Ni_X$ with $x =$ 34.2–35.5 at.%. It is concluded that the HPT at 298 K of the initial quenched and annealed alloys decreases the homogeneity of short-range order due to its destruction by mobile dislocations. An increase in the deformation temperature from room temperature to 473 K makes the ordering less pronounced, whereas homogeneous short-range ordering accelerates at 573 K. This process is characterized by an increase in the neighbourhood number of Fe and Ni atoms and the appearance of a stage similar to radiation-accelerated ordering. The ordering accelerated by warm deformation is similar to that accelerated by radiation. It is caused by the continuous generation of non-equilibrium mobile point defects upon treatment [25].

In Fe-7.6 wt% Si alloy, it is found that the spatially-directed anisotropic short-range order is present after cooling due to low-diffusion mobility of atoms, leading to the anisotropy of magnetic properties [26]. Effects of chromium additions on the microstructure, ordered phase and magnetic properties of the hot-rolled and warm-rolled Fe-6.5 wt% Si alloys were investigated by Chen et al. [27]. It was concluded that the chromium addition reduced the grain size of the alloy effectively, and the dislocation density was reduced with the increasing of the chromium content [27].

7.2.2 Microstructure Characterization of SRO in Metallic Materials

The determination of the atomic configuration of metallic materials has been a long-term challenge in materials science. Several characterization techniques and analytic methods can be employed to describe SRO, for example, the pair distribution function (PDF) method. In PDF analysis, a pair correlation represents the probability of finding atoms as a function of the distance from an average centre atom. The structure factor is directly related to the partial pair distribution functions in reciprocal space via Fourier transformation [28]. The variable in the reciprocal space corresponds to the magnitude of the diffraction vector and can be determined by X-ray scattering experiments.

The SRO characterization techniques include: (i) X-ray/neutron diffraction and scattering: experimentally, the measured scattering intensity can be related to the total structure factor. Recently, new XRD techniques is proposed and developed to uncover more details regarding the local packing symmetry and 3D structure of amorphous materials, such as the X-ray cross correlation analysis (XCCA) [29], and ankylography [30]. (ii) X-ray absorption fine structure (XAFS): It includes the extended X-ray absorption fine structure (EXAFS) and X-ray absorption near edge structure (XANES), is a commonly used method for structural characterization [31]. (iii) Fluctuation electron microscopy (FEM): FEM is an electron microscopy technique that is especially sensitive to medium-range (of the order of 1 nm) structural features [32]. By sweeping across the sample using nanometre-sized X-ray beam and measuring the fluctuation of the scattering intensity, one can assess the degree of ordering over medium range in the material. (iv) High-resolution transmission electron microscopy (HRTEM): It enables direct observation of atomic-level structure, but only as a projected 2-D image. (v) The nuclear magnetic resonance (NMR) technique: It has been applied to study the local symmetry around certain atoms [33].

A combination of small angle neutron scattering (SANS) and atom probe tomography (APT) is employed to investigate the effect of Mn additions on the size, composition and number density of the copper-enriched precipitates in Fe-0.80Cu (at.%) and Fe-0.78Cu1.05Mn (at.%) model alloys [34]. The SANS measurements in a mm^3 volume show that larger and less numerous precipitates are formed in the Fe-Cu than the Fe-Cu-Mn alloy [34]. High entropy alloys (HEA), medium entropy alloys (MEA) and Ti-based alloys are reported to form the SRO and can be analysed by the sophisticated techniques. Castany et al. [15] used transmission electron microscopy (TEM) to quantitatively characterization the SRO and its effects on the deformation micro-mechanism in a Ti-6Al-4V alloy. The local order of a Ti-6Al-4V alloy is identified to be SRO because of the absence of superlattice diffraction spots, which are associated with α_2 (Ti_3Al) precipitates and because of the formation of single pairs of mobile dislocations, which are a signature of SRO [15].

Santodonato et al. reported the new alloy-design strategy of combining segregation, precipitation, chemical ordering and spinodal decomposition to process the HEA with the stoichiometry of $Al_{1.3}CoCrCuFeNi$ [35]. Two types of submicron

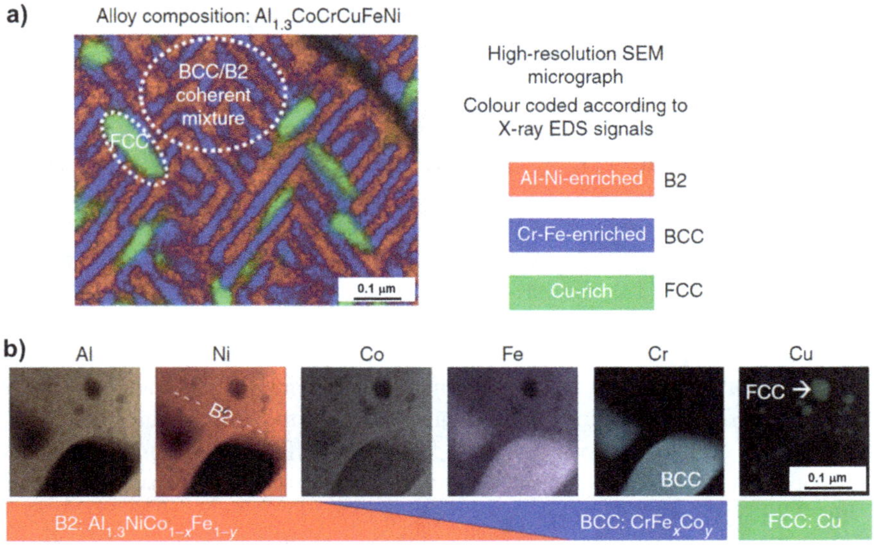

Fig. 7.1 Microscopy and elemental analyses of $Al_{1.3}$CoCrCuFeNi alloy: **a** The high-resolution SEM micrograph is colour coded according to EDS, highlighting the elements segregation; **b** The abundance of each individual element is shown in a series of TEM/EDS images of the same sample region [35]

Cu-rich precipitates, rod shaped (Fig. 7.1a) and spherical (Fig. 7.1b), were characterized by TEM—energy dispersive X-ray spectrum (EDS). The rods are dispersed throughout the dendrites (that is, the bcc/B2 matrix), but the spheres are found only in the spinodal B2 phase. Given that the total volume of Cu in the material is 16%, it may be inferred that the volume fraction of the Cu nano-precipitates is on the order of 1% (from above, the Cu-rich rods and interdendritic regions account for 15%). Taking all of the above Cu-rich fcc phases into account, the adjusted volume phase fractions of the alloy are 50% B2, 34% bcc and 16% fcc structures [35].

Ding et al. revealed the nature of local chemical order and established its significant relationship to the intrinsic and extrinsic stacking fault energy in CrCoNi medium-entropy solid solution alloys [36]. The research results highlight the significance of local chemical ordering in determination of mechanical properties, which can improve the science-based design and optimization of new high-entropy alloy systems with specifically desired combinations of macroscale mechanical properties [36]. Simulations employing a density function theory (DFT)-based lattice Monte Carlo (MC) approach are proven to be effective in qualitatively correlating with the experimental measurements. These replicated samples lead to values close to those for the as-assigned supercells. This result provides validation for our calculation of stacking fault energy (SFE) using the replicated configurations, which forms the basis for the studies, described in the next section below, of the effects of chemical SRO on SFE [36].

7.2.3 *Influence of SRO on the Mechanical Properties in Metallic Materials*

SRO influences various mechanical properties of materials, e.g. deformation behaviours [15, 37–40], fatigue [41], creep resistance [7, 42], magnetic and electrical properties [26], and stress corrosion cracking susceptibility [16]. Large amount of research has been carried out on SRO in the last decades, i.e. the binary alloys of Cu-Mn [37], Au_4Mn [43], αCu-Al [44], Ni-20%Cr [38], Ti-6 wt%Al [7]; the ternary alloys of Ti-6Al-4V [15], the $Al_{50}Ge_{40}Mn_{10}$ amorphous alloy [45] and the alloy $Co_{58}Ni_{10}Fe_5Si_{11}B_{16}$ [46]. It is suggested that a good strength-ductility balance in Cu-Mn alloys can be achieved by increasing degree of SRO [37]. As the Mn content increases from 5 to 20 at.%, the dominating deformation microstructures of Cu-Mn alloy evolves from dislocation cells to planar-like slip structures, and further planar slip structures and dislocation twinnings. The overall effect of SRO on the mechanical properties and the relevant microstructures are schematically displayed shown in Fig. 7.2 [37].

SRO affects the stacking fault energy which controls the plasticity mechanisms in fcc-Fe [47]. Substitutional and interstitial impurities may strongly affect the SFE and the plasticity mechanism. Figure 7.3 shows the intrinsic SFE and unstable stacking fault energies (USFE) in fcc-Fe as dependent on Mn concentration and Al concentration [47]. It is suggested that the Mn, Al and C in fcc Fe have an influence on the SFE only when they are located within a few atomic layers near a stacking fault [47].

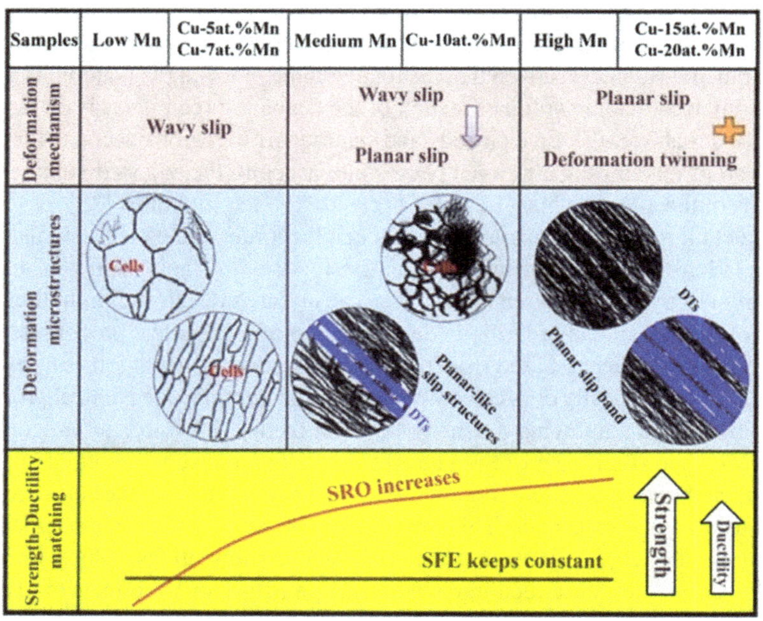

Fig. 7.2 Effect of the SRO on the mechanical properties and microstructure [37]

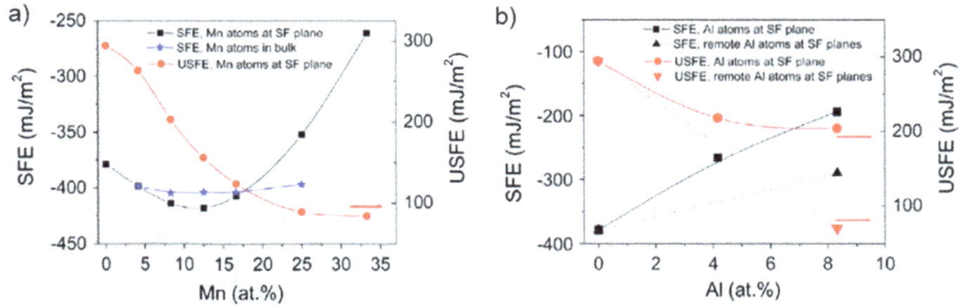

Fig. 7.3 The intrinsic stacking fault energies (SFE) and unstable stacking fault energies (USFE) in fcc-Fe as dependent on **a** Mn concentration and **b** Al concentration [47]

7.3 Applications

7.3.1 Study on a High Entropy Alloy by High Energy Synchrotron X-Ray Diffraction and SANS

The concept of HEA is one of the most recent developments in materials science [48–53], which shows extraordinary mechanical properties such as high strength [53], outstanding high-temperature strength [49, 53–62], good ductility [63, 64], good toughness [65], corrosion resistance [66–69], wear and fatigue resistance [70, 71]. There are four core factors predominating the mechanical properties in HEAs [50, 56]: (i) the high entropy effect; (ii) the sluggish diffusion effect [56, 72]; (iii) the severe lattice distortion effect; (iv) the "cocktail effect", which refers to the formation of a synergistic mixture where the end result is unpredictable and greater than the sum of the parts. Among four critical factors, the lattice distortion effect is crucial to understand the amplification mechanisms of HEAs, especially single-phase HEAs. The occupation of the same crystallographic sites by the different sized atoms in the solid solution leads to a local distortion of the lattice [56]. Lattice distortion can be experimentally measured by the diffraction and scattering method, e.g. high energy synchrotron X-ray and neutron radiation [51, 56, 73–80], which allow high penetration and/or high resolution measurements instead of laboratory X-ray diffraction. The HEAs based on CoCrFeNi system shows good mechanical properties with further additions of Al [49, 50, 81–84], Cu [49, 50, 85], Si [86], Mn [81] and Mo [87]. However, the quaternary CoCrFeNi-based HEA has not been intensively investigated, especially the lattice distortion effect that contributes significantly to the strength enhancement. To systematically study the lattice distortion in HEAs, adequate knowledge of two types of distortions is supposed to be acquired. The first type is long-range lattice distortion, which leads to strain fields in the HEAs. The second type is local lattice distortion at the atomic level. This type of lattice distortion does not broaden the Bragg peaks but can be analysed using the PDF method, which determines the real-space distribution of atoms in the lattice of HEAs [88].

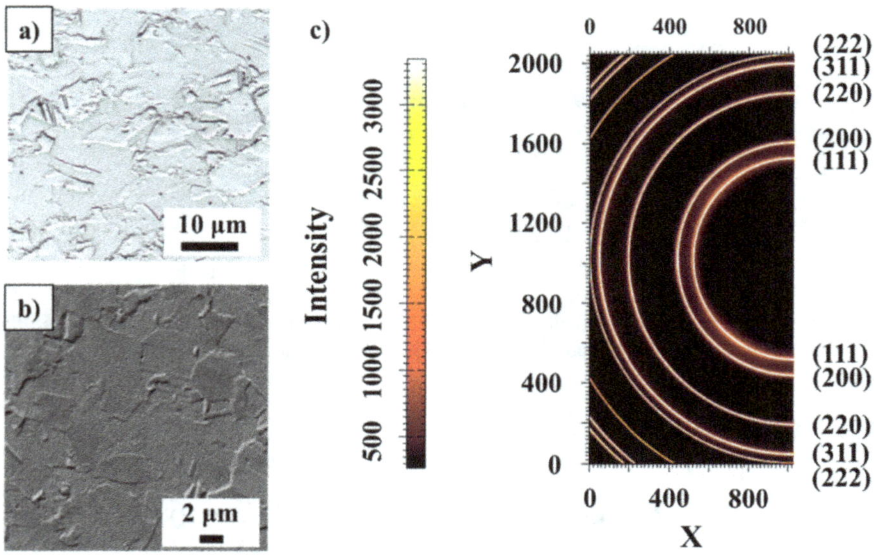

Fig. 7.4 **a** LOM; **b** SEM and **c** 2D SYXRD pattern of the CoCrFeNi HEA [88]

In this subchapter, the microstructure of a quaternary CoCrFeNi HEA was investigated using high energy synchrotron X-ray diffraction (SYXRD) and SANS [88]. In the as-forged condition, the CoCrFeNi HEA reveals purely face-centered cubic (fcc) crystal structure. The dendritic structure remained from primary solidification was eliminated by the hot forging process. And the average grain size obtained is about 11.6 μm. The results show that a pronounced lattice distortion is observed in CoCrFeNi-HEA compared to pure Ni, which is believed to contribute to the solid solution strengthening of the HEA. The CoCrFeNi HEA has an excellent yield strength ($R_{p0.2} = 710$ MPa at room temperature and $R_{p0.2} = 660$ MPa at 500 °C). The results of the small-angle neutron scattering experiment also show the local ordering effect in the fcc solid solution of CoCrFeNi-HEA both at room temperature and at 500 °C after tensile deformation [88].

Figure 7.4a, b show the microstructure of the CoCrFeNi HEA at the as-forged state [88]. Figure 7.4c exhibits the 2D SYXRD pattern of the CoCrFeNi HEA, where 5 full diffraction rings are indexed, i.e. (111), (200), (220), (311) and (222). The converted intensity versus 2-theta curve of the CoCrFeNi HEA is shown in Fig. 7.5a. For comparison, the 1D diffraction peaks of pure Ni are shown in Fig. 7.5b [88]. Compared to pure Ni, the HEA Bragg peaks are slightly shifted to the lower 2-theta angle, indicating the larger austenite lattice parameter in the CoCrFeNi HEA. The lattice parameter of the CoCrFeNi HEA is measured as 3.573 Å. The lattice is distorted by 1.23% compared to pure Ni with a lattice parameter of 3.529 ± 0.001 Å [88].

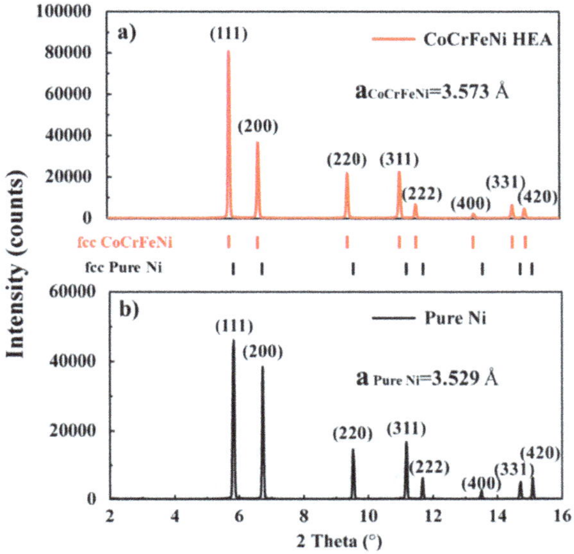

Fig. 7.5 1D synchrotron diffraction peaks of the **a** CoCrFeNi HEA and **b** pure Ni [88]

For a detailed analysis on the peak shift and lattice distortion, the (111) and (200) planes of the CoCrFeNi HEA and the pure Ni at a 2 theta range of 5.4°–7.0° are further plotted in Fig. 7.6a [88]. Additional impurities are introduced into the CoCrFeNi HEA lattice compared to pure Ni, which contributes to the lattice distortion in the HEA and the peak shift in the 1D diffraction pattern in the SYXRD measurement. For (111) plane, the $\Delta\theta$ value is 0.084° and $\Delta\theta$ for the (200) plane is 0.085° [88]. According to the Smallman and Westmacott's equation [89], the dislocation density can be calculated proportionally to the average microstrain $\langle\varepsilon^2\rangle^{1/2}$ and inversely proportional to the average crystallite size D:

$$\rho = \frac{3\sqrt{2\pi}\left(\varepsilon^2\right)^{1/2}}{Db} \tag{7.1}$$

where b denotes the Burgers vector. With further analysis of microstrain and crystallite size by Rietveld refinement, the dislocation density in CoCrFeNi HEA is calculated to be $8.064 \times 10^{14}/\text{m}^2$ [88].

The nuclear scattering curves of the SANS measurement of the CoCrFeNi HEA are plotted in Fig. 7.6b [88]. The scatter curves were fitted with a spherical model assuming the contrast to be 2×10^{10} cm^{-2}. The ordering effect in CoCrFeNi HEA was observed by SANS measurement as shown in Fig. 7.6b, where the Q range of the ordering is indicated by arrows. It is evident that the ordering occurs both at room temperature and at 500 °C after deformation [88]. Compared to the order at room temperature, the mean size of ordering phase increases at 500 °C in CoCrFeNi-HEA. In both cases, at low Q, the scattering indicating larger structures with sizes in the (m-range) shows up as Porod-like Q^{-4} scattering behavior and this is probably due to

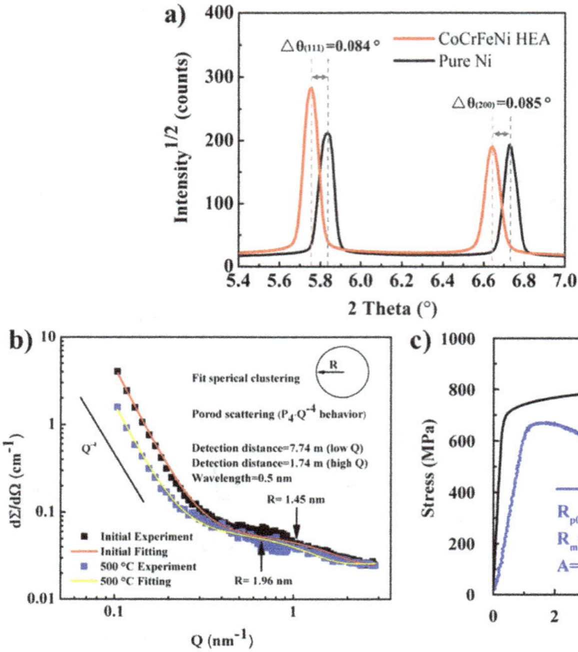

Fig. 7.6 **a** (111) and (200) synchrotron diffraction peaks of CoCrFeNi HEA and pure Ni; **b** SANS scattering curves showing the ordering effect in CoCrFeNi HEA; **c** Flow curves of the CoCrFeNi HEA determined by tensile test at room temperature and at 500 °C [88]

the fcc grains of the matrix. Figure 7.6c shows the engineering stress–strain curves of the CoCrFeNi HEA at room temperature and 500 °C. The single fcc phase CoCrFeNi HEA reveals the yield strength ($R_{p0.2}$) of 710 MPa, an ultimate tensile strength (R_m) of 803 MPa and maintains a total elongation (A) of 11.3%. At 500 °C, the yield strength of the CoCrFeNi HEA reaches 660 MPa and ultimate tensile strength is around 673 MPa [88]. CoCrFeNi HEA has excellent yield strength in the as-forged condition both at room temperature and at high temperature, in comparison with the similar systems [80, 90–92].

Briefly, a pronounced lattice distortion phenomenon and a local ordering effect were observed in the HEA and their contribution to the mechanical properties was discussed. In the as-forged condition, the single-phase CoCrFeNi HEA achieved an excellent yield strength ($R_{p0.2}$) of 710 MPa at room temperature and 660 MPa at 500 °C [88]. The sophisticated characterization tools, SYXRD and SANS, were employed to study the lattice distortion and ordering effect, thanks to their high-resolution detection properties in the nano/atomic range and statistical measurement in large volumes in mm^3. The interesting observation of the local order effect in the HEAs is particularly worth mentioning. Since local ordering could greatly increase the yield strength and ductility of the alloys, but the mechanism is still unclear, it is of great importance to gain further insights on how this phenomenon can be controlled and applied to obtain improved mechanical properties in HEAs [88]. In the future, the nature and evolution of the order and its role in the deformation mechanisms in HEAs will be further explored to enable a new design of HEAs with outstanding mechanical properties [88].

7.3.2 Experimental Observation of Kappa Phase Formation Sequences by In-Situ Synchrotron Diffraction

The addition of significant amounts of Al to steel provides a dual benefit of reducing the overall density of the steel and increasing its strength due to the formation of intermetallic phases. The latter contribution to strength requires a profound knowledge of the phase transformation in the precipitation of intermetallic phases. The Fe_3AlC or termed as kappa phase can be formed in high-Al alloys either in a face-centred cubic (fcc) austenite or body-centred cubic (bcc) ferrite matrix. The research of precipitation behaviour has been conducted, inspired by the beneficial contribution of the kappa phase to the mechanical properties of fully austenitic high-manganese steels [93, 94]. The kappa phase exhibits a $L'1_2$-crystal structure, with Fe and Al arranged on simple cubic sublattices, and C occupying the center of the lattice structure, forming a perovskite-like structure. It has been proposed that spinodal decomposition occurs during the phase transformation from austenite (γ) to the kappa phase [95, 96]. The decisive difference between disordered austenite and ordered configurations lies in a transition from fcc to simple cubic structures and thus an additional superlattice reflection. {100}-superreflection corresponding to a single crystal lattice structure are identified in Fe-Mn-Al-C steels [97, 98].

A possible formation sequence of the kappa phase is shown schematically in Fig. 7.7 (see red arrows). Possible reaction sequences for first-order transformations are extensively discussed by Soffa and Laughlin [99]. A disordered fcc state is cooled from high temperatures and complete long-range ordering over the range of a few atomic distances occurs via diffusion. This ordered structure then separates into a disordered fcc phase and an $L'1_2$-structured kappa phase, possibly by spinodal decomposition. The aforementioned reaction sequence was reported by Tan et al. [100] in Ni-base superalloys. It should be noted that both the previously formed ordered phase and the last formed kappa phase have an $L'1_2$ crystal structure, with the difference being in the Al and C concentration (Fig. 7.7) [101]. Therefore, the two types of configurations are referred to as C-poor and C-rich kappa phase, respectively. The lack of dynamic monitoring of the kappa phase formation brings uncertainties in transformation sequence and its underlying kinetics [101].

In this subchapter, the first in-situ experimental observation of kappa phase formation by in-situ synchrotron diffraction during heating of a high-Al-alloyed lightweight

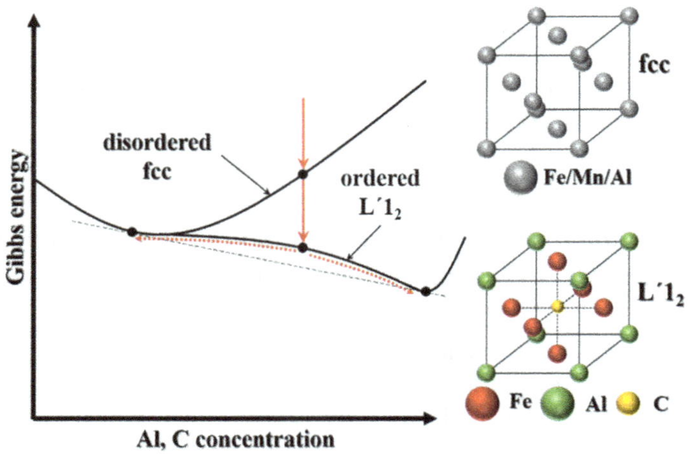

Fig. 7.7 Schematical illustration of Gibbs energy curves in dependence of the Al- and C-concentration and the crystal structures of participating phases [101]

steel was performed [101]. In-situ annealing experiments using synchrotron diffraction enables the deconvolution of {100}-superlattice peaks. The peak deconvolution is the strong experimental in-situ techniques for determining the transformation sequence during Fe_3AlC-phase formation. An exhaustive analysis of the lattice parameter evolution and thermal expansion behaviour was carried out to verify the allocation of phases by comparison with theoretical computations and accessible experimental data [101].

The chemical composition of the investigated duplex steel Fe-10Al-7Mn-6Ni-1C (in wt%) is provided in Table 7.1 [101]. Homogenization was carried out at 1200 °C for 5 h before hot open die forging and water quenching. Solution treatment was performed under an Ar atmosphere at 1200 °C for 2 h, followed by water quenching. Synchrotron X-ray diffraction was carried out in the high-resolution powder diffraction beamline P02.1 at DESY, Hamburg. In-situ heating was conducted in a ceramic heating oven. Heating rate of 15 K/min was employed until an isothermal holding temperature of 800 °C was reached [101]. Data conversion and subsequent Rietveld refinement were performed using Fit2D [102] and open source software "Materials Analysis Using Diffraction" (MAUD) [103].

The diffraction profile of the solution annealed sample is displayed in Fig. 7.8a [101]. The ordered phases of the $L'1_2$- and B2-type are present in the zoom-in view, where the superlattice reflections corresponding to the $\{100\}_{L'1_2}$- and $\{110\}_{L'1_2}$-reflections as well as the $\{100\}_{B2}$-peak are indexed. Figure 7.8b displays the evolution

Table 7.1 Chemical composition of the multi-phase alloy [101]

Fe-10Al-7Mn-6Ni-1C	Al	Mn	Ni	C	Si	P	S	N
wt%	10.1	6.8	6.1	0.95	0.025	0.012	0.011	0.012

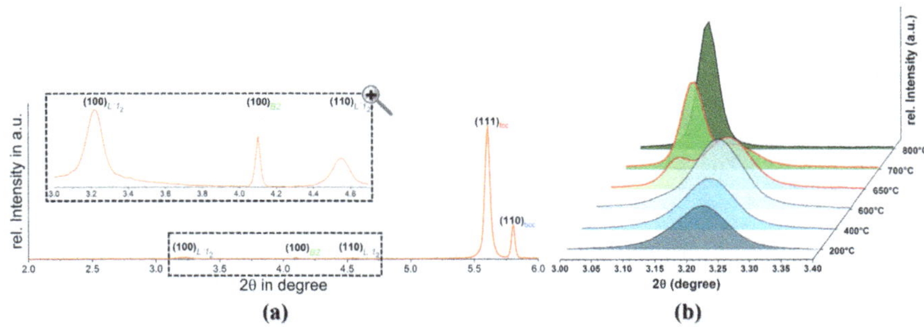

Fig. 7.8 **a** Synchrotron diffraction profile of the duplex alloy in the as-quenched condition with enlarged view of superreflexes corresponding to $L'1_2$- and B2-structures; **b** Evolution of $\{100\}_{L'1_2}$-superlattice reflection during heating from 200 °C to isothermal holding temperature of 800 °C [101]

of the $\{100\}_{L'1_2}$-superlattice reflection during heating from 200 to 800 °C [101]. The peak intensity increases up to 600 °C, indicating an increase in volume fraction. At the same time, the peak position continuously shifts to smaller angles due to the modification to the lattice parameter. The significant shift in the peak position of the secondary reflection is attributed to the proceeding phase transformation. Further at 700 °C, most of the superlattice reflection transforms into the newly formed configuration. The overall conversion is complete when 750 °C is reached and thus takes place in approx. 5 min [101].

Figure 7.9 depicts the evolution of the lattice parameter during aging process. The lattice parameter for C-depleted $L'1_2$ structure increases from 3.672 to 3.735 Å. The ordering effect occurs from 650 °C and leads to an increase of lattice parameter to an overall mean of 3.821 ± 0.039 Å [101]. Figure 7.7 schematically illustrate the transformation sequence, showing that fcc- and kappa phase can reach their respective equilibrium state. The matrix-particle coherence between the fcc and kappa phases forces the phases to maintain an equal lattice parameter [101]. At the equilibrium state, the kappa phase exhibits a larger lattice parameter compared with fcc phase, increasing further with the elevated C content.

The coherent kappa phase thus has a lower C content than the equilibrium kappa phase [101], leading to the pronounced peak separation of the superlattice reflection in XRD spectrum. Both C-rich and C-lean kappa phases give rise to superlattice reflections, while enrichment of the C-lean kappa phase is expected to induce a prompt increase in the lattice parameter [101]. The formation of the C-lean kappa phase is attributed to the fast growth kinetics associated with the high-C diffusivity at elevated temperatures. The measurement of the C-lean kappa phase is considered to be suppressed, since the completion of ordering and the subsequent phase separation to a C-rich kappa phase appears plausible under certain cooling conditions [101].

The kappa phase frequently exhibits an off-stoichiometric concentration of Al and C [104]. The scatter in chemical composition of the kappa phase complicates the analysis of lattice parameter. C enrichment of fcc and its related ordered structure

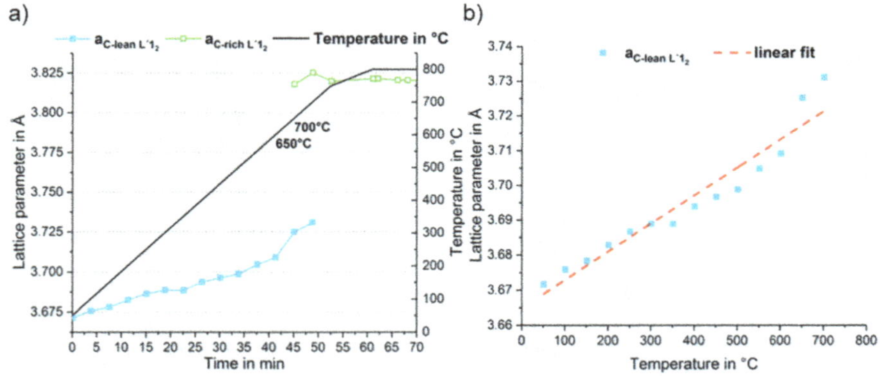

Fig. 7.9 **a** Evolution of the lattice parameter during heating of the C-lean and C-rich $L'1_2$-structure and **b** Thermal expansion of the C-lean $L'1_2$-phase [101]

is expected to increase the lattice parameter [105, 106]. The C-enrichment of the kappa phase with respect to its parent fcc-phase was previously investigated by atom probe tomography measurements [105, 107]. The elevated C concentration induces an increase in lattice parameter by approximately 0.09 Å at 650 °C. The thermal expansion of the ordered $L'1_2$-phase is investigated in detail, as shown in Fig. 7.9b [101]. The C-lean $L'1_2$-lattice parameter increases near-linearly with temperature and the thermal expansion is approximated empirically to be:

$$a_{(L'1_2)}^{(C\text{-}lean)}(T) = (3.66505 \pm 0.00285)\,\text{Å} + (8.06418 * 10^{(-5)} \pm 0.66981 * 10^{(-5)})\,\text{Å}/K * T \tag{7.2}$$

The room temperature lattice parameter of the C-rich kappa phase reveals a range from 3.72 to 3.78 Å according to the experimental measurements [105, 106, 108, 109] and theoretical calculations [104, 110]. Assuming the similar thermal expansion of the ordered structures, the room temperature lattice parameter of the kappa phase $a_{L'1_2}^{C-rich}(RT)$ is estimated to be 3.756 Å by recalculating $a_{L'1_2}^{C-rich}$ at 800 °C based on the thermal expansion coefficient of the C-lean kappa phase in Eq. 7.2 [101].

Maugis et al. [111] determined the room temperature lattice parameter of the Fe_3Al-$L1_2$ and Fe_3AlC-$L'1_2$ as 3.65 and 3.75 Å, using ab initio calculations. Similar calculations of the theoretical lattice parameter of $L1_2$-Fe_3Al and $L'1_2$-Fe_3AlC showed a small deviation of 0.04–0.10 Å [110, 112]. The consistency of theoretical calculations and experimental observations in lattice parameter confirms the transformation sequence from disordered fcc phase to an ordered C-lean kappa phase, and subsequently followed by a phase separation into disordered fcc and C-rich kappa phase [101]. In summary, in-situ synchrotron diffraction identifies the kappa phase formation sequence, where the separation of diffraction peaks is correlated to the evolution of a C-rich kappa phase from an ordered C-lean κ-precursor. The assignment of the phases was validated by comparing the lattice parameters with theoretical calculations and experimental data. The observed transformation sequence indicates a prior ordering of the disordered fcc-phase into a long-range ordered C-lean κ-precursor, which subsequently transform into a disordered fcc- and C-rich κ-phase [101].

7.3.3 Local Deformation and Mn-C Short-Range Ordering in a High-Mn Fe-18Mn-0.6C Steel

High-Mn steels (HMnS), as one of the promising candidates of 2nd generation of advanced high strength steel, provide an extraordinary combination of strength and ductility [113–116]. A unique feature of such steels is the significant serration behaviour upon stretching under certain conditions of strain rate and temperature, known as the Portevin-Le Chatelier (PLC) effect [113, 116]. The serrations in flow curves are traditionally classified into three types: A, B and C [117–120], based on their shapes. Type A serrations were observed in Fe-18Mn-0.6C steel and PLC band properties were analysed by Chen et al. [113], Kim et al. [116] and Hong et al. [121]. However, the correlation of serrated flow curves and PLC bands has not been comprehensively understood. During deformation, plastic strain is localized in the form of bands that move in various ways along the specimen gauge. The nucleation and movement of PLC bands involve the collective movement of a large number of dislocations [122–124]. Since stress localization causes deterioration of inherent structural properties and surface quality of materials [125], understanding the PLC effect is crucial for the effective application of high-Mn steels. The localized deformation in high-Mn steels is explained by the interaction between Mn-C pairs and dislocations. The re-orientation of interstitial C in the Mn-C complex may segregate at the core of moving dislocations [117, 126]. Based on the thermodynamics [127], the combination of thermoelectric power and internal friction measurements [128], Monte Carlo simulation [129], an attractive interaction between Mn and C is proved to form octahedral Mn-C clusters in the Fe-Mn-C alloy system [130]. Mn-C clusters are defined as Mn-C SRO, since arrangement of atoms in Mn-C clusters is over a distance comparable to interatomic distances [130].

In the present work [130], to study the influence of Mn-C SRO on the localized deformation in high-Mn steels, digital image correlation (DIC) assisted uniaxial tensile tests were conducted on Fe-18Mn-0.6C steel for the simultaneous investigation of the localized deformation, namely, serrated flow and PLC bands. SANS was utilized to detect the presence of Mn-C SRO and quantitatively analyse its size and number density. The chemical composition of the investigated cold rolled austenitic HMnS is Fe-18Mn-0.6C (in wt%) [130]. The PLC band velocity v_B was calculated by $v_B = l_B/t_B$, here l_B is the displacement of band movement and t_B is the time for it. To study the impacts of strain on the interactions of Mn-C SRO and dislocations during plastic deformation, the tensile tests were carried out at a constant strain rate of 1×10^{-3} s^{-1} at room temperature. Besides the fractured sample with the total

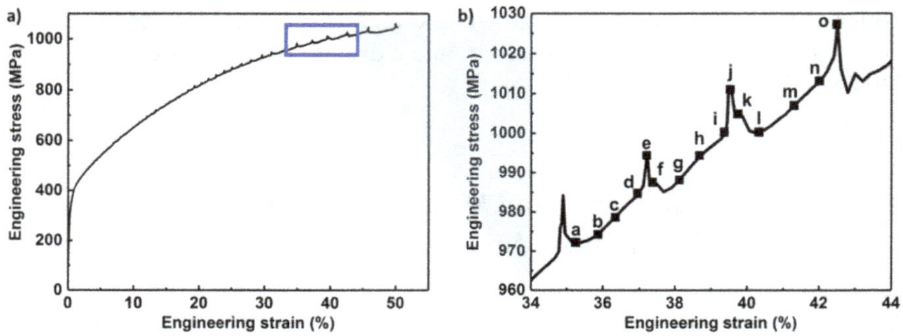

Fig. 7.10 **a** Engineering stress–strain curves of annealed Fe-18Mn-0.6C steel and **b** The serrated flow between 34 and 44% strain [130]

elongation of 45%, three interrupted tensile tests with the engineering strains of 15, 25, 35% were imposed for 2 min annealed samples [130].

The engineering stress-engineering strain curve of Fe-18Mn-0.6C steel after recrystallization is illustrated in Fig. 7.10a, showing a yield stress of 340 MPa, an ultimate tensile strength of 1060 MPa while maintaining a total elongation of 50% [130]. Stepped stress jumps are observed, in which plateau areas are separated by a slight increase in stress (serration). The magnified view of the serrated flow curve in the range of engineering strain from 34 to 44% is shown in Fig. 7.10b. Spikes i.e. e, j and o, denote serration points. As the strain increased, the distance between two serration points increased and the stress rise at the prongs increased [130].

The PLC bands are visualized by DIC-assisted imaging analysis and those corresponding to point a–o in Fig. 7.10b are displayed in Fig. 7.11a [130]. The PLC bands are visualized by localized strain rate contour. The PLC band nucleated firstly in the middle region and propagated to one end of the sample, followed by another PLC band forming at the same site but moving towards the opposite end. The distribution of localized strain rate shows that the strain of the specimen is concentrated inside the PLC bands (in red), while the remaining part outside the PLC band remains almost no strain (in blue). The PLC bands in stage e, j and o are outside the gauge length, which corresponds to the serrations in Fig. 7.10b. The strain distribution of the PLC bands in Fig. 7.11b is represented by the strain contour, ranging from 33% (blue) to 43% (red). The localized strain is more concentrated than the overall engineering strain [130]. At stage a, for instance, the strain is localized in the middle region at PLC band, while remaining part maintains a relatively lower strain; the local strain in the vicinity of the PLC band reaches over 40% and the engineering strain is around 35%. In the area where PLC band has already swiped, the local strains are increased accordingly [130].

Figure 7.12b shows the serrated flow curve and the corresponding PLC band at different stain levels (~15, ~25, ~35, ~45%) visualized by DIC technique [130]. The morphology of serrated flow at ~15% strain is different from those at ~25, ~35 and ~45% strains since the serrations are less obvious (Point e of Fig. 7.12a. The

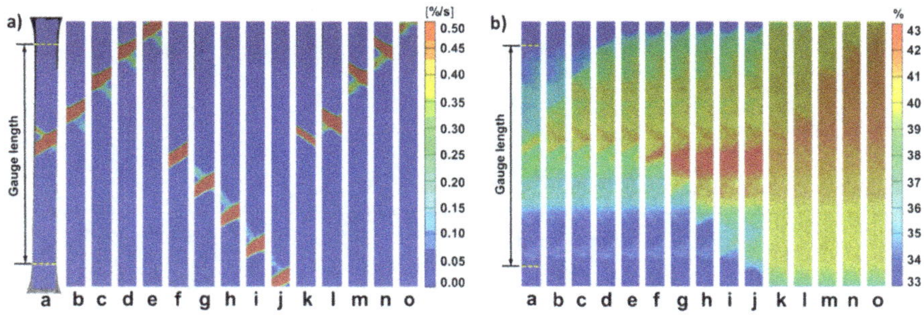

Fig. 7.11 **a** PLC bands and **b** Local strain distribution in Fig. 7.10b [130]

corresponding PLC band of Point a–f in Fig. 7.12a is shown in Fig. 7.12b. The PLC band nucleates in the middle region and propagates to the top end. The strain rate along the tensile direction is plotted, shown in Fig. 7.12c [130]. The local strain rate within the PLC band reaches the value of 0.6% s^{-1}. The strain distribution of stage a–f is shown in Fig. 7.12d, where the strain is mainly localized in the PLC band [130]. The localized strain denotes 17% s^{-1}, higher than the average engineering strain (~15% s^{-1}). A similar trend is observed at strain of ~25, ~35 and ~45%. The serrations at ~25% (Fig. 7.12e), ~35% (Fig. 7.12i) and ~45% (Fig. 7.12m) are larger than that of strain ~15% (Fig. 7.12a) [130]. The PLC bands at the evaluated strains are shown in Fig. 7.12f (~25%), Fig. 7.12j (~35%) and Fig. 7.12n (~45%), whose widths are smaller than that of strain at ~15% (Fig. 7.12b). The PLC bands nucleate at the same place—the middle of the gauge section and propagates to one end of it. The local strain rate reaches 1% s^{-1} at strain of ~25% (Fig. 7.12g), 1.6% s^{-1} at strain of ~35% (Fig. 7.12k) and 2% s^{-1} at strain of ~45% (Fig. 7.12o) respectively. The local strain within the PLC band amounts to 28% at strain of ~25% (Fig. 7.12h), 40% at strain of ~35% (Fig. 7.12l) and 50% at strain of ~45% (Fig. 7.12p) respectively [130].

The PLC band expanding velocity v_B decreases rapidly as the engineering strain increases, as shown in Fig. 7.13 [130]. It is observed that v_B decreased linearly with the increase in engineering strain. The band velocity drops from ~4 mm/s (~strain of 15%) to ~1 mm/s (~strain of 48%) [130]. The scatter of the value is greater for strains less than 30%, while the value for strains greater than 30% is more narrowly distributed along the fitting curve (black straight line) [130].

SANS was then used to characterize the Mn-C SRO in the deformed samples. Sections upstream and downstream of the PLC bandpass were compared to investigate the influence of location on localized strain behaviour of Mn-C SRO [130]. The sections after PLC band passed by were selected to investigate the impacts of strain and strain rate on Mn-C SRO. Due to the inhomogeneity of the microstructure, the SANS analysis can provide statistical results with an average homogeneous size distribution of the particles. The particle in SANS measurement is assumed to be a cluster form of Mn-C pairs (Mn_6C) during the deformation process [130]. Figure 7.14a shows the microscopic differential cross-section $d\sigma/d\Omega$ (in cm^{-1}) as a

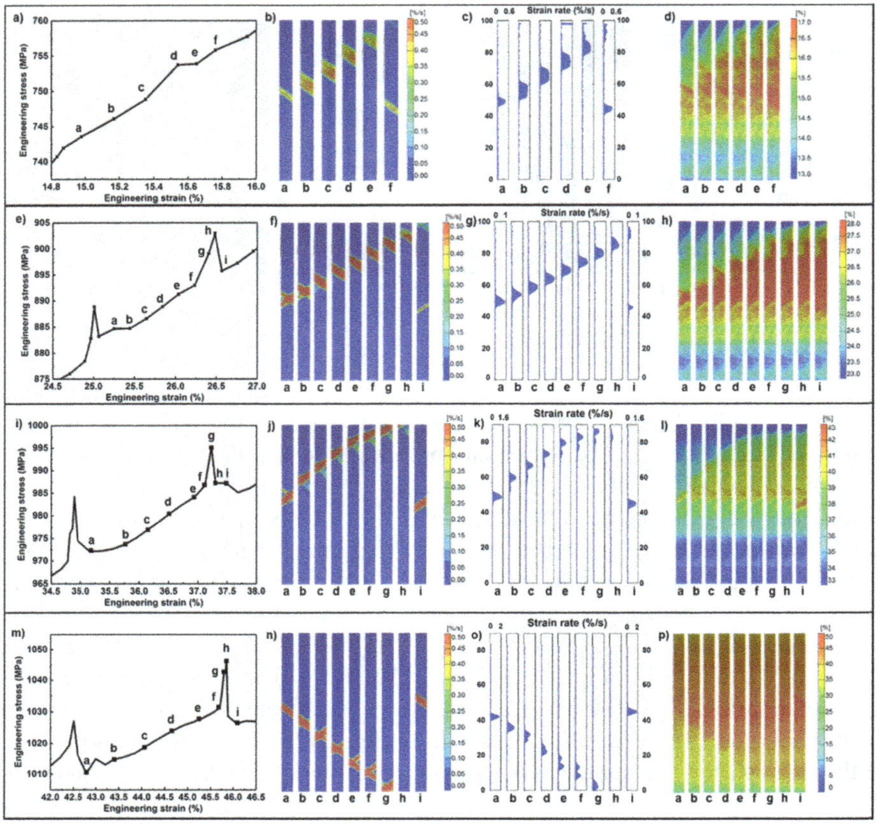

Fig. 7.12 Serrated slow, strain rate contour, strain rate distribution and strain distribution along tensile direction at strain ~15%: **a–d**; ~25%: **e–h**; ~35%: **i–l**; and ~45%: **m–p** [130]

Fig. 7.13 PLC band velocity during tensile test [130]

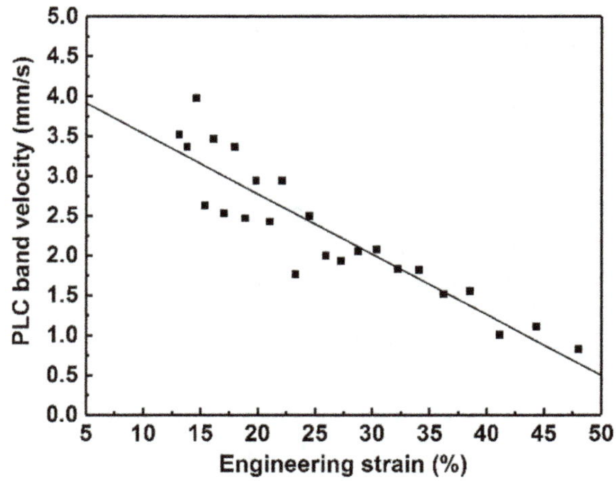

function of q range according to SANS in a log–log plot, obtained from samples at different strain levels, i.e. 15, 25, 35, and 45% respectively [130]. The microscopic differential cross-section is found to decrease with the increase in scattering vector q (in nm^{-1}). The q ranges can be divided into three types, i.e., low ($q \leq 0.4$), middle ($0.4 \leq q \leq 1$), and high ($1 \leq q$). The middle q range is focused in the present study for the analysis of particles with the size of 3–60 nm. The humps in the middle q range in Fig. 7.14a reveals the presence of Mn-C SRO [130]. With the increase in strain from 15 to 45%, the hump in middle q range become more obvious.

The shift in q-area indicates that the greater the strain experienced by the specimens, the smaller the Mn-C SRO in the tensile specimen. Figure 7.14b shows the evolution of the size distribution of the Mn-C SRO with increasing strain. The mean radius and number density obtained from Fig. 7.14b is plotted in Fig. 7.14c [130]. The radius of the Mn-C SRO decreases while the number density increases dramatically with the increasing strain. The Mn-C SRO has an average radius of 2.18 nm after the recrystallization annealing and decreases to 1.87 nm (~15% strain), 1.68 nm (~25% strain), 1.46 nm (~35% strain) and finally to 0.88 nm (~45% strain) in the deformed samples [130]. Furthermore, the number density of Mn-C SRO increases from 1.05×10^{15} cm^{-3} (as-annealed) to 1.31×10^{15} cm^{-3} (~15% strain), 2.56×10^{15} cm^{-3} (~25% strain), 6.40×10^{15} cm^{-3} (~35% strain) and 1.86×10^{17} cm^{-3}

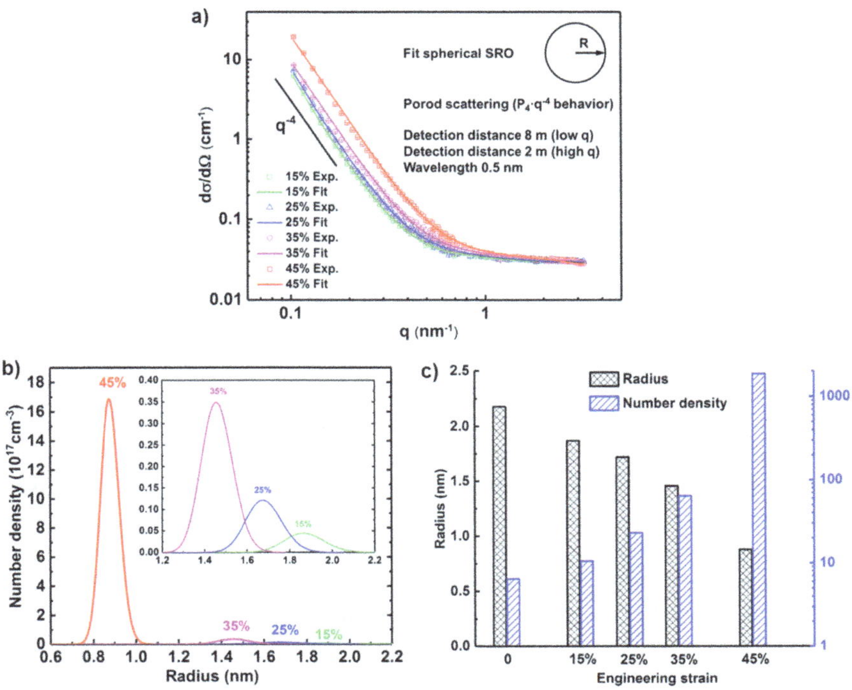

Fig. 7.14 **a** SANS scattering curves; **b** Number density size distribution and **c** Mean radius and number density of Mn-C SRO in deformed samples at strain from 15 to 45% [130]

(~45% strain) [130]. When a PLC band propagates during plastic deformation, the behaviours of Mn-C SRO in different locations, i.e. before and after the band passes by, are of interest.

Figure 7.15 shows the number density size distribution of Mn-C SRO at different conditions [130]. The position before PLC band passage refers to A and that after band passed by indicates B for simplification in the text. The mean size of Mn-C SRO at site A is larger than that at site B at strains of 15, 25, 35 and 45%. The number density shows an increasing trend when comparing the value at site A with that at site B. The number density at site A is lower than at site B for all strain states [130]. The serration behaviours generally begin at a critical applied strain with a fine and short shape, indicating that dynamic strain aging (DSA) does not usually occur immediately after the onset of straining. This is explained by the fact that sufficient vacancies must be created first during tension to facilitate the solute diffusion to dislocations for DSA [131]. In contrast, Cuddy and Leslie [132] argued the formation of vacancies during deformation cannot predominate the occurrence of serrations. Based on the DIC analysis in Fig. 7.12, the nucleation of each PLC band is correlated with a load drop, while the load gradually increases during the propagation of each PLC band, accounting for the formation of plateau region between two serration points on the flow curves [130]. The inhomogeneity in the microstructure of the samples might be controlling factor. The nucleation sites of PLC bands are not constant, i.e. at either end of the gauge or in the middle of the sample [116, 117, 133]. As the PLC band propagates beyond the gauge length and dies down, the load suddenly increases and forms a serration before the next PLC band nucleates, which could be attributed to the geometrical factor [134]. Since the plastic deformation is localized in the experiments performed, the critical force required depends on the deforming specimen cross-section. In order to achieve plastic deformation at a constant strain rate, a larger force should be applied to the sample at the region beyond gauge length [130]. The serration amplitude $\Delta\sigma$ and the serration interval $\Delta\varepsilon_{BS}$ are found to elevate with increasing strain for each experiment [130].

PLC band velocity decreases linearly with increasing strain, and fracture occurs as PLC band velocity approaches near zero [130]. Korbel et al. [135] found that increasing the mobile dislocation density decreases PLC band velocity and increases band strain. This implies that when a PLC band passes through the sample, the material is strengthened and a large amount of barriers to plastic flow are created, suppressing the movement of successive band [136]. Upon PLC starts to propagate, the local strain rate inside the PLC band is at least one order higher than that outside the band. The strain increment can thus be concentrated in the PLC band during the propagation process [130]. Once the propagation of the PLC band velocity is retarded to almost zero, the strain will be localized, leading to uniform plastic deformation and final fracture [137]. In the recrystallization annealed Fe-18Mn-0.6C samples, the model-based analysis of SANS patterns reveals the presence of nano-sized Mn-C clusters, indicating Mn-C SRO formation during annealing [130]. Thermodynamic data [127] and internal friction measurements [128] demonstrate an attractive interaction between Mn and C in austenite due to the reduced activity of C with the presence of Mn. Indeed, the thermochemical driving force should lead to a Mn enrichment at

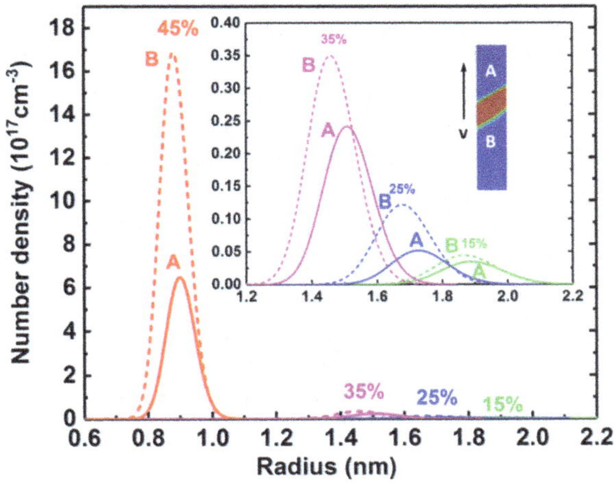

Fig. 7.15 Mn-C SRO in location A (before PLC band passes by) and B (after PLC band passed by) in deformed samples with strain from 15 to 45% [130]

the immediate vicinity of the carbon atom and the energy difference indicates a strong tendency for carbon-induced SRO [129, 138]. The ab initio simulation indicates that carbon atoms repel each other and avoid the nearest occupations at the nearest interstitial sites. In annealed Fe-18Mn-0.6C steel, a Mn_6C octahedron is energetically favourable, while the C-C repulsive interaction can counteract the Mn-C attraction, resulting in the formation of Mn_6C units that are evenly distributed rather than tightly packed to be grouped or to keep minimum distances mutually [139]. Thus, the longer annealing time leads to the decrease in size and increase in number density of Mn-C SRO [130].

The Mn-C SRO shows a decreasing tendency in the mean radius and an increasing tendency in the number density with increasing strain in the plastic deformation. This agrees with the recent SANS study [115] on the Mn-C cluster formation in Fe-18Mn-0.6C steel, showing the increased cluster volume fraction and number density during plastic deformation, which could be attributed to two dominant mechanisms [130]. On the one hand, Mn-C SRO is sheared by the dislocations during plastic deformation. Pronounced SRO or short-range clusters (SRC) have been suggested to be the primary reason for planar slip [140]. SRO or SRC may be dissolved through shearing interaction with dislocations, resulting in the glide softening. In precipitation-hardened alloys, the finely dispersed precipitates with an atomic order also exhibit the similar behaviour [130]. In Al-alloyed HMnS the nano-sized kappa phases are considered to be shearable due to coherency and tiny size, acting as weak obstacles for dislocation motion [141]. Similarly, the Mn-C SRO of a few nanometres in size could be sheared by the dislocations during plastic deformation, leading to an increase in number density and a decrease in size, contributing to the superior combination of tensile strength and ductility [130]. One the other hand, the strain-induced new-born Mn-C SRO can also account for the change in SRO characteristics. Kandarpa and Spretnak [142] concluded that the internal friction peak in austenitic manganese steel originated from stress-induced ordering of carbon atoms, which is owing to the distortion of octahedral sites induced by neighbouring Mn atoms [130]. Such ordering become more stable within dislocation cores than in the elastically strained lattice. The influence of deformation prior to aging on the precipitation behaviour and mechanical enhancement has been reported in the literature [143]. In HMnS, the DSA can be observed at room temperature by the rearrangement of C atoms of at point defects and their interaction with the dislocations and stacking fault regions [113, 116, 117]. Deformation reduces the activation energy of C rearrangement and eases the DSA at room temperature [117]. The deformation-induced defects and reduction in activation energy required for C rearrangement could lead to the evolution of Mn-C-SRO upon deformation [130].

In the present work, the influence of Mn-C SRO on localized deformation in Fe-18Mn-0.6C steel were investigated using a combined means of SANS and DIC analysis. With the increasing strain in uniaxial tensile tests, the mean radius of Mn-C SRO decreases while the number density of Mn-C SRO increases [130]. The Mn-C SRO at the point after the passing PLC band exhibits a smaller SRO size with a higher number density compared to that in the portion before the PLC passing by. With large applied strain (>15%), PLC bands form steadily and continuously. On the flow curve, each segment (plateau) between two nearby serration points corresponds to the nucleation and propagation of a PLC band within the gauge length. The interlocking corresponds to the beginning/disappearance of the PLC band outside the measuring length [130]. With increasing strain, the velocity of PLC bands decreases. The longer recrystallization annealing time leads to an accelerated PLC band velocity. The PLC band velocity is correlated with the high strain rate applied in the tensile tests. The strain is distributed heterogeneously and is localized within the PLC bands during plastic deformation. In the section after passing the PLC band, the local strain is greater than in the section without PLC band passing by [130].

7.4 Summary

The newly developed alloys are prone to be heavily alloyed, e.g. high-Mn steels and high-entropy alloys, in order to meet the application-determined property requirements. In these alloy systems, the SRO formation is inevitable. By means of a combined method of ab initio calculations, in-situ high energy SYXRD and SANS, the interstitials-contained SRO is for the first time observed. It is of significant importance to understand the SRO formation in nanostructured metals and to utilize it as an effective strengthening approach.

References

1. Cohen J, Fine M (1962) Some aspects of short-range order. J Phys Radium 23:749–762
2. Gladman T (1999) Precipitation hardening in metals. Mater Sci Technol 15:30–36

3. Kim S, Kim H, Kim N (2015) Brittle intermetallic compound makes ultrastrong low-density steel with large ductility. Nature 518:77–79
4. Jiang S, Wang H, Wu Y, Liu X, Chen H, Yao M, Gault B, Ponge D, Raabe D, Hirata A, Chen M, Wang Y, Lu Z (2017) Ultrastrong steel via minimal lattice misfit and high-density nanoprecipitation. Nature 544:460–464
5. Anthony L, Fultz B (1995) Effects of early transition metal solutes on the $D0_3$–B2 critical temperature of Fe_3Al. Acta Metall Mater 43:3885–3891
6. Van Tendeloo G (1976) Short range order considerations and development of long range order in different Ni-Mo alloys. Mater Sci Eng 26:209–220
7. Neeraj T, Mills M (2001) Short-range order (SRO) and its effect on the primary creep behavior of a Ti-6 wt.% Al alloy. Mater Sci Eng A 319–321:415–419
8. Yang Z, Zhang L, Chisholm M, Zhou X, Ye H, Pennycook S (2018) Precipitation of binary quasicrystals along dislocations. Nat Commun 9:261
9. Ringer S, Hono K, Sakurai T, Polmear I (1997) Cluster hardening in an aged Al-Cu-Mg alloy. Scr Mater 36:517–521
10. Starink M, Wang S (2009) The thermodynamics of and strengthening due to co-clusters: general theory and application to the case of Al-Cu-Mg alloys. Acta Mater 57:2376–2389
11. Chen Y, Gao N, Sha G, Ringer S, Starink M (2015) Strengthening of an Al-Cu-Mg alloy processed by high-pressure torsion due to clusters, defects and defect-cluster complexes. Mater Sci Eng A 627:10–20
12. Marceau R, Sha G, Ferragut R, Dupasquier A, Ringer S (2010) Solute clustering in Al-Cu-Mg alloys during the early stages of elevated temperature ageing. Acta Mater 58:4923–4939
13. Murayama M, Hono K (1999) Pre-precipitate clusters and precipitation processes in Al-Mg-Si alloys. Acta Mater 47:1537–1548
14. Aruga Y, Kozuka M, Takaki Y, Sato T (2016) Effects of natural aging after pre-aging on clustering and bake-hardening behavior in an Al-Mg-Si alloy. Scr Mater 116:82–86
15. Castany P, Pettinari-Sturmel F, Douin J, Coujou A (2017) TEM quantitative characterization of short-range order and its effects on the deformation micromechanisms in a Ti-6Al-4V alloy. Mater Sci Eng A 680:85–91
16. Kim Y, Maeng W, Kim S (2015) Effect of short-range ordering on stress corrosion cracking susceptibility of Alloy 600 studied by electron and neutron diffraction. Acta Mater 83:507–515
17. Kang J, Ingendahl T, von Appen J, Dronskowski R, Bleck W (2014) Impact of short-range ordering on yield strength of high manganese austenitic steels. Mater Sci Eng A 614:122–128
18. Sevsek S, Bleck W (2018) *Ab initio*-based modelling of the yield strength in high-manganese steels. Metals 8:34
19. Morris D, Munoz-Morris M (2011) Recent developments toward the application of iron aluminides in fossil fuel technologies. Adv Eng Mater 13:43–47
20. Witusiewicz V, Bondar A, Hecht U, Velikanova T (2011) Phase equilibria in binary and ternary systems with chemical and magnetic ordering. J Phase Equilib Diff 32:329–349
21. Thomas H (1951) Über Widerstandslegierungen. Z Phys 129:219–232
22. Lin Y, Chou C (1993) $D0_3$-B2-α phase transition in an Fe-Mn-Al-C weldment. Scr Metall Mater 28:1261–1266
23. Shabashov V, Kozlov K, Sagaradze V, Nikolaev A, Semyonkin V, Voronin V (2018) Short-range order clustering in bcc Fe-Mn alloys induced by severe plastic deformation. Philos Mag 98:560–570
24. Kozlov K, Shabashov V, Zamatovskii A, Sagaradze V, Lyashkov K (2018) Atomic ordering in a low-concentrated Fe-Cr alloy upon severe plastic deformation. Phys Metals Metallogr 119:1093–1100
25. Shabashov V, Kozlov K, Zamatovskii A, Lyashkov K, Sagaradze V, Danilov S (2019) Short-range atomic ordering accelerated by severe plastic deformation in fcc invar Fe-Ni alloys. Struct Phase Trans Diff 120:686–693
26. Chernenkov Y, Ershov N, Lukshina V, Fedorov V, Sokolov B (2007) An X-ray diffraction study of the short-range ordering in the soft-magnetic Fe-Si alloys with induced magnetic anisotropy. Phys B 396:220–230

27. Chen W, Liu J, Cheng Z, Lin X, Zhu J (2015) Effect of chromium on microstructure, ordered phase and magnetic properties of Fe-6.5 wt.% Si alloy. Mater Today Proc 2:314–318
28. Egami T, Billinge S (2003) Underneath the Bragg peaks: structural analysis of complex materials. In: Cahn R (ed) Pergamon materials series. Oxford, Elsevier
29. Wochne P, Gutt C, Autenrieth T, Demmer T, Bugaev V, Diaz Ortiz A, Duri A, Zontone F, Gruebel G, Dosch H (2009) X-ray cross correlation analysis uncovers hidden local symmetries in disordered matter. Proc Natl Acad Sci (PANS) 106:11511–11514
30. Raines K, Salha S, Sandberg R, Jiang H, Rodríguez J, Fahimian B, Kapteyn H, Du J, Miao J (2010) Three-dimensional structure determination from a single view. Nature 463:214–217
31. Rehr J, Albers R (2000) Theoretical approaches to X-ray absorption fine structure. Rev Mod Phys 72:621–654
32. Treacy M, Gibson J, Fan L, Paterson D, Mcnulty I (2005) Fluctuation microscopy: a probe of medium range order. Rep Prog Phys 68:2899–2944
33. Xi X, Sandor M, Liu Y, Wang W, Wu Y (2009) Structural changes induced by microalloying in $Cu_{46}Zr_{47-x}Al_7Gd_x$ metallic glasses. Scr Mater 61:967–969
34. Miller M, Wirth B, Odette G (2003) Precipitation in neutron-irradiated Fe-Cu and Fe-Cu-Mn model alloys: a comparison of APT and SANS data. Mater Sci Eng A 353:133–139
35. Santodonato L, Zhang Y, Feygenson M, Parish C, Gao M, Weber R, Neuefeind J, Tang Z, Liaw P (2015) Deviation from high-entropy configurations in the atomic distributions of a multi-principal-element alloy. Nat Commun 6:5964
36. Ding J, Yu Q, Asta M, Ritchie R (2018) Tunable stacking fault energies by tailoring local chemical order in CrCoNi medium-entropy alloys. PNAS 115:8919–8924
37. Han D, Wang Z, Yan Y, Shi F, Li X (2017) A good strength-ductility match in Cu-Mn alloys with high stacking fault energies: determinant effect of short-range ordering. Scr Mater 133:59–64
38. Dudova N, Kaibyshev R (2010) Short-range ordering and mechanical properties of a Ni-20% Cr alloy. Int J Phys Conf Ser 240:012081
39. Karaman I, Sehitoglu H, Maier H, Chumlyakov Y (2001) Competing mechanisms and modeling of deformation in austenitic stainless steel single crystals with and without nitrogen. Acta Mater 49:3919–3933
40. Turk C, Leitner H, Kellezi G, Clemens H, Gan W, Staron P, Primig S (2016) Impact of the B2 ordering behavior on the mechanical properties of a FeCoMo alloy. Mater Sci Eng A 662:511–518
41. Kang J, Zhang F, Long X, Lv B (2014) Cyclic deformation and fatigue behaviors of Hadfield manganese steel. Mater Sci Eng A 591:59–68
42. Zhao B, Fan J, Liu Y, Zhao L, Dong X, Sun F, Zhang L (2015) Formation of $L1_2$-ordered precipitation in an alumina-forming austenitic stainless steel via Cu addition and its contribution to creep/rupture resistance. Scr Mater 109:64–67
43. Suzuki H, Harada J, Nakashima T, Adachi K (1982) Short-range ordering and ferromagnetic properties of disordered Au_4Mn alloy. Acta Crystallogr Sect A 38:522–529
44. Tomokiyo Y, Kuwano N, Eguchi T (1975) Short range ordering in deformed α Cu-Al alloys. Trans Jpn Inst Metals 16:489–499
45. Inoue A, Bizen Y, Kimura H, Yamamoto M, Tsai A, Masumoto T (1987) Development of compositional short-range ordering in an $Al_{50}Ge_{40}Mn_{10}$ amorphous alloy upon annealing. J Mater Sci Lett Mater Sci Eng A 6:811–814
46. Dong X, Fernengel W, Kronmüller H (1982) Annealing effects and short-range ordering in the non-magnetostrictive amorphous alloy $Co_{58}Ni_{10}Fe_5Si_{11}B_{16}$. Appl Phys A 28:103–107
47. Medvedeva N, Park M, Van Aken D, Medvedeva J (2014) First-principles study of Mn, Al and C distribution and their effect on stacking fault energies in fcc Fe. J Alloy Compd 582:475–482
48. Huang K (1996) A study on the multicomponent alloy systems containing equal-mole elements. Master thesis, National Tsinghua University
49. Yeh J, Chen S, Lin S, Gan J, Chin T, Shun T, Tsau C, Chang S (2004) Nanostructured high-entropy alloys with multiple principal elements: novel alloy design concepts and outcomes. Adv Eng Mater 6:299–303

50. Yeh J (2006) Recent progress in high-entropy alloys. Ann Chim Sci Mat 31:633–648
51. Salishchev G, Tikhonovsky M, Shaysultanov D, Stepanov N, Kuznetsov A, Kolodiy I, Tortika A, Senkov O (2014) Effect of Mn and V on structure and mechanical properties of high-entropy alloys based on CoCrFeNi system. J Alloy Compd 591:11–21
52. Youssef K, Zaddach A, Niu C, Irving D, Koch C (2015) A novel low-density, high-hardness, high-entropy alloy with close-packed single-phase nanocrystalline structures. Mater Res Lett 3:95–99
53. He J, Wang H, Huang H, Xu X, Chen M, Wu Y, Liu X, Nieh T, An K, Lu Z (2016) A precipitation-hardened high-entropy alloy with outstanding tensile properties. Acta Mater 102:187–196
54. Senkov O, Scott J, Senkova S, Miracle D, Woodward C (2011) Microstructure and room temperature properties of a high-entropy TaNbHfZrTi alloy. J Alloy Compd 509:6043–6048
55. Juan C, Tsai M, Tsai C, Lin C, Wang W, Yang C, Chen S, Lin S, Yeh J (2015) Enhanced mechanical properties of HfMoTaTiZr and HfMoNbTaTiZr refractory high-entropy alloys. Intermetallics 62:76–83
56. Tsai M, Yeh J (2014) High-entropy alloys: a critical review. Mater Res Lett 2:107–123
57. Zhang Y, Yang X, Liaw P (2012) Alloy design and properties optimization of high-entropy alloys. JOM 64:830–838
58. Tang Z, Gao M, Diao H, Yang T, Liu J, Zuo T, Zhang Y, Lu Z, Cheng Y, Zhang Y, Dahmen K, Liaw P, Egami T (2013) Aluminium alloying effects on lattice types, microstructures, and mechanical behavior of high-entropy alloys systems. JOM 65:1848–1858
59. Zuo T, Ren S, Liaw P, Zhang Y (2013) Processing effects on the magnetic and mechanical properties of FeCoNiAl 0.2 Si 0.2 high entropy alloy. Int J Miner Metall Mater 20:549–555
60. Yang X, Zhang Y, Liaw P (2012) Microstructure and compressive properties of NbTiVTaAlx high entropy alloys. Procedia Eng 36:292–298
61. Senkov O, Wilks G, Miracle D, Chuang C, Liaw P (2010) Refractory high-entropy alloys. Intermetallics 18:1758–1765
62. Antonaglia J, Xie X, Tang Z, Tsai C, Qiao J, Zhang Y, Laktionova M, Tabachnikova E, Yeh J, Senkov O, Gao M, Uhl J, Liaw P, Dahmen K (2014) Temperature effects on deformation and serration behavior of high-entropy alloys (HEAs). JOM 66:2002–2008
63. Zhang Y, Zuo T, Cheng Y, Liaw P (2013) High-entropy alloys with high saturation magnetization, electrical resistivity, and malleability. Sci Rep 3:1455
64. Laktionova M, Tabchnikova E, Tang Z, Liaw P (2013) Mechanical properties of the high-entropy alloy Ag0.5CoCrCuFeNi at temperatures of 4.2–300 K. Low Temp Phys 39:630–632
65. Gludovatz B, Hohenwarter A, Catoor D, Chang E, George E, Ritchie R (2014) A fracture-resistant high-entropy alloy for cryogenic applications. Science 345:1153–1158
66. Tang Z, Huang L, He W, Liaw P (2014) Alloying and processing effects on the aqueous corrosion behavior of high-entropy alloys. Entropy 16:895–911
67. Lee C, Chen Y, Hsu C, Yeh J, Shih H (2008) Enhancing pitting corrosion resistance of AlxCrFe1.5MnNi0.5 high-entropy alloys by anodic treatment in sulfuric acid. Thin Solid Films 517:1301–1305
68. Qiu X, Zhang Y, He L, Liu C (2013) Microstructure and corrosion resistance of AlCrFeCuCo high entropy alloy. J Alloys Compd 549:195–199
69. Kai W, Li C, Cheng F, Chu K, Huang R, Tsay L, Kai J (2016) The oxidation behavior of an equimolar FeCoNiCrMn high-entropy alloy at 950 °C in various oxygen-containing atmospheres. Corros Sci 108:209–214
70. Chuang M, Tsai M, Wang W, Lin S, Yeh J (2011) Microstructure and wear behavior of $Al_xCo_{1.5}CrFeNi_{1.5}Ti_y$ high-entropy alloys. Acta Mater 59:6308–6317
71. Hemphill M, Yuan T, Wang G, Yeh J, Tsai C, Chuang A, Liaw P (2012) Fatigue behavior of Al0.5CoCrCuFeNi high entropy alloys. Acta Mater 60:5723–5734
72. Cheng K, Lai C, Lin S, Yeh J (2006) Recent progress in multi-element alloy and nitride coatings sputtered from high-entropy alloy. Ann Chim 31:723–736
73. Tong C, Chen Y, Yeh J, Lin S, Chen S, Shun T, Tsau C, Chang S (2005) Microstructure characterization of AlxCoCrCuFeNi high-entropy alloy system with multiprincipal elements. Metall Mater Trans A 36:881–893

74. Chou H, Chang Y, Chen S, Yeh J (2009) Microstructure, thermophysical and electrical properties in AlxCoCrFeNi (0≤x≤2) high-entropy alloys. Mater Sci Eng B 163:184–189
75. Yeh J, Chen Y, Lin S, Chen S (2007) High-entropy alloys—a new era of exploitation. Mater Sci Forum 560:1–9
76. Wang F, Zhang Y (2008) Effect of Co addition on crystal structure and mechanical properties of Ti0.5CrFeNiAlCo high entropy alloy. Mater Sci Eng A 496:214–216
77. Wang F, Zhang Y, Chen G (2009) Atomic packing efficiency and phase transition in a high entropy alloy. J Alloys Compd 478:321–324
78. Guo W, Dmowski W, Noh J, Rack P, Liaw P, Egami T (2013) Local atomic structure of a high-entropy alloy: an x-ray and neutron scattering study. Metall Mater Trans A 44:1994–1997
79. Yao M, Pradeep K, Tasan C, Raabe D (2014) A novel, single phase, non-equiatomic FeMn-NiCoCr high-entropy alloy with exceptional phase stability and tensile ductility. Scr Mater 72:5–8
80. Tsai C, Chen Y, Tsai M, Yeh J, Shun T, Chen S (2009) Deformation and annealing behaviors of high-entropy alloy Al0.5CoCrCuFeNi. J Alloys Compd 486:427–435
81. Zhou Y, Zhang Y, Wang Y, Chen G (2007) Microstructure and compressive properties of multicomponent Alx(TiVCrMnFeCoNiCu)100–x high-entropy alloys. Mater Sci Eng A 454:260–265
82. Zang K, Fu Z, Zhang J, Wang W, Wang H, Wang Y, Zhang Q, Shi J (2009) Microstructure and mechanical properties of CoCrFeNiTiAlx high-entropy alloys. Mater Sci Eng A 508:214–219
83. Kao Y, Chen T, Chen S, Yeh J (2009) Microstructure and mechanical property of as-cast, -homogenized, and -deformed AlxCoCrFeNi (0≤x≤2) high-entropy alloys. J Alloys Compd 488:57–64
84. Li C, Li J, Zhao M, Jiang Q (2010) Effect of aluminum contents on microstructure and properties of AlxCoCrFeNi alloys. J Alloys Compd 504:515–518
85. Zhou Y, Zhang Y, Wang F, Wang Y, Chen G (2008) Effect of Cu addition on the microstructure and mechanical properties of AlCoCrFeNiTi0.5 solid-solution alloy. J Alloys Compd 466:201–204
86. Zhu J, Fu H, Zhang H, Wang A, Li H, Hu Z (2010) Synthesis and properties of multiprincipal component AlCoCrFeNiSix alloys. Mater Sci Eng A 527:7210–7214
87. Zhu J, Fu H, Zhang H, Wang A, Li H, Hu Z (2010) Microstructures and compressive properties of multicomponent AlCoCrFeNiMox alloys. Mater Sci Eng A 527:6975–6979
88. Song W, Radulescu A, Liu L, Bleck W (2017) Study on a high entropy alloy by high energy synchrotron x-ray diffraction and small angle neutron scattering. Steel Res Int 88:1700079
89. Smallman R, Westmacott K (1957) Stacking faults in face-centred cubic metals and alloys. Philos Mag 2:669–683
90. Otto F, Dlouhý A, Somsen C, Bei H, Eggeler G, George E (2013) The influences of temperature and microstructure on the tensile properties of a CoCrFeMnNi high-entropy alloy. Acta Mater 61:5743–5755
91. Deng Y, Tasan C, Pradeep K, Springer H, Kostka A, Raabe D (2015) Design of a twinning-induced plasticity high entropy alloy. Acta Mater 94:124–133
92. Li Z, Pradeep K, Deng Y, Raabe D, Tasan C (2016) Metastable high-entropy dual-phase alloys overcome the strength-ductility trade-off. Nature 534:227
93. Chin K, Lee H, Kwak J, Kang J, Lee B (2010) Thermodynamic calculation on the stability of (Fe, Mn)3AlC carbide in high aluminum steels. J Alloys Compd 505:217–223
94. Song W, Zhang W, von Appen J, Dronskowski R, Bleck W (2015) k-phase formation in Fe-Mn-Al-C austenitic steels. Steel Res Int 86:1161–1169
95. Choo W, Kim J, Yoon J (1997) Microstructural change in austenitic Fe-30.0 wt.%Mn-7.8 wt.%Al-1.3wt.%C initiated by spinodal decomposition and its influence on mechanical properties. Acta Mater 45:4877–4885
96. Cheng W, Cheng C, Hsu C, Laughlin D (2015) Phase transformation of the L12 phase to kappa carbide after spinodal decomposition and ordering in an Fe-C-Mn-Al austenitic steel. Mater Sci Eng A 642:128–135

97. Ha M, Koo J, Lee J, Hwang S, Park K (2013) Tensile deformation of a low density Fe-27Mn-12Al-0.8C duplex steel in association with ordered phases at ambient temperature. Mater Sci Eng A 586:276–283
98. Wang C, Hwang C, Chao C, Liu T (2007) Phase transitions in an Fe-9Al-30Mn-2.0C alloy. Scr Mater 57:809–812
99. Soffa W, Laughlin D (1989) Decomposition and ordering processes involving thermodynamically first-order order → disorder transformations. Acta Metall 37:3019–3028
100. Tan X, Mangelinck D, Perrin-Pellegrino C, Rougier L, Gandin C, Jacot A, Ponsen D, Jaquet V (2014) Spinodal decomposition mechanism of γ' precipitation in a single crystal Ni-based superalloy. Metall Mater Trans A 45:4725–4730
101. Drouven C, Hallstedt B, Song W, Bleck W (2019) Experimental observation of κ-phase formation sequences by *in-situ* synchrotron diffraction. Mater Lett 241:111–114
102. Hammersley A (1997) FIT2D: an introduction and overview. Eur Synchrotron Radiat Facil Intern Rep 68:58
103. Ferrari M, Lutterotti L (1994) Method for the simultaneous determination of anisotropic residual stresses and texture by x-ray diffraction. J Appl Phys 76:7246–7255
104. Yao M, Dey P, Seol J, Choi P, Herbig M, Marceau R, Hickel T, Neugebauer J, Raabe D (2016) Combined atom probe tomography and density functional theory investigation of the Al off-stoichiometry of κ-carbides in an austenitic Fe-Mn-Al-C low density steel. Acta Mater 106:229–238
105. Kim C, Kwon S, Lee B, Moon J, Park S, Lee J, Hong H (2016) Atomistic study of nano-sized κ-carbide formation and its interaction with dislocations in a cast Si added FeMnAlC lightweight steel. Mater Sci Eng A 673:108–113
106. Bartlett L, Aken D, Medvedeva J, Isheim D, Medvedeva N, Song K (2014) An atom probe study of kappa carbide precipitation and the effect of silicon addition. Metall Mater Trans A 45:2421–2435
107. Seol J, Raabe D, Choi P, Park H, Kwak J, Park C (2013) Direct evidence for the formation of ordered carbides in a ferrite-based low-density Fe-Mn-Al-C alloy studied by transmission electron microscopy and atom probe tomography. Scr Mater 68:348–353
108. Sato K, Tagawa K, Inoue Y (1989) Spinodal decomposition and mechanical properties of an austenitic Fe-30 wt.%Mn-9 wt.%Al-0.9 wt.%C alloy. Mater Sci Eng A 111:45–50
109. Sato K, Tagawa K, Inoue Y (1990) Modulated structure and magnetic properties of age-hardenable Fe-Mn-Al-C alloys. Metall Trans A 21:5–11
110. Connétable D, Maugis P (2008) First principle calculations of the κ-Fe_3AlC perovskite and iron-aluminium intermetallics. Intermetallics 16:345–352
111. Maugis P, Lacaze J, Besson R, Morillo J (2006) *Ab initio* calculations of phase stabilities in the Fe−Al−C system and CALPHAD-Type assessment of the iron-rich corner. Metall Mater Trans A 37:3397–3401
112. Reddy B, Deevi S (2002) Local interactions of carbon in FeAl alloys. Mater Sci Eng A 329–331:395–401
113. Chen L, Kim H, Kim S, De Cooman B (2007) Localized deformation due to Portevin-LeChatelier effect in 18Mn-0.6C TWIP austenitic steel. ISIJ Int 47:1804–1812
114. Kim J, Chen L, Kim H, Kim S, Kim G, Estrin Y, De Cooman B (2009) Strain rate sensitivity of C-alloyed, high-Mn, twinning-induced plasticity steel. Steel Res Int 80:493–498
115. Kang M, Shin E, Woo W, Lee Y (2014) Small-angle neutron scattering analysis of Mn-C clusters in high-manganese 18Mn-0.6C steel. Mater Char 96:40–45
116. Kim J, Chen L, Kim H, Kim S, Estrin Y, De Cooman B (2009) On the tensile behavior of high-manganese twinning-induced plasticity steel. Metall Mater Trans A 40:3147
117. Lee S, Kim J, Kane S, De Cooman B (2011) On the origin of dynamic strain aging in twinning-induced plasticity steels. Acta Mater 59:6809–6819
118. Brindley B, Worthington P (1970) Yield-point phenomena in substitutional alloys. Metall Rev 15:101–114
119. Pink E, Grinberg A (1981) Serrated flow in a ferritic stainless steel. Mater Sci Eng A 51:1–8

120. Wang W, Wu D, Shah S, Chen R, Lou C (2016) The mechanism of critical strain and serration type of the serrated flow in Mg-Nd-Zn alloy. Mater Sci Eng A 649:214–221
121. Hong S, Shin SY, Lee J, Ahn DH, Kim HS, Kim SK, Chin KG, Lee S (2014) Serration phenomena occurring during tensile tests of three high-manganese twinning-induced plasticity (TWIP) steels. Metall Mater Trans A 45:633–646
122. Kubin L, Ananthakrishna G, Fressengeas C (2002) Comment on "Portevin-Le Chatelier effect." Phys Rev E 65:053501
123. Hähner P (1996) On the physics of the Portevin-Le Chatelier effect part 2: from microscopic to macroscopic behaviour. Mater Sci Eng A 207:216–223
124. Hähner P (1996) On the physics of the Portevin-Le Châtelier effect part 1: the statistics of dynamic strain ageing. Mater Sci Eng A 207:208–215
125. Yilmaz A (2011) The Portevin-Le Chatelier effect: a review of experimental findings. Sci Technol Adv Mater 12
126. Dastur Y, Leslie W (1981) Mechanism of work hardening in Hadfield manganese steel. Metall Mater Trans A 12:749–759
127. Chipman J, Brush E (1968) The activity of carbon in alloyed austenite at 1000 C. Trans Metall Soc AIME 242:35–41
128. Massardier V, Merlin J, Le Patezour E, Soler M (2005) Mn-C interaction in Fe-C-Mn steels: study by thermoelectric power and internal friction. Metall Mater Trans A 36:1745–1755
129. Owen W, Grujicic M (1998) Strain aging of austenitic Hadfield manganese steel. Acta Mater 47:111–126
130. Song W, Houston JE (2018) Local deformation and Mn-C short-range ordering in a high-Mn Fe-18Mn-0.6C steel. Metals 8:292
131. Cottrell A (1953) A note on the Portevin-Le Chatelier effect. Phil Mag 44:829–832
132. Cuddy L, Leslie W (1972) Some aspects of serrated yielding in substitutional solid solutions of iron. Acta Metall 20:1157–1167
133. Zavattieri P, Savic V, Hector L, Fekete J, Tong W, Xuan Y (2009) Spatio-temporal characteristics of the Portevin-Le Châtelier effect in austenitic steel with twinning-induced plasticity. Int J Plast 25:2298–2330
134. Zdunek J, Spychalski W, Mizera J, Kurzydłowski K (2007) The influence of specimens geometry on the PLC effect in Al-Mg-Mn (5182) alloy. Mater Char 58:46–50
135. Korbel A, Zasadzinski J, Sieklucka Z (1976) A new approach to the Portevin-LeCatelier effect. Acta Metall 24:919–923
136. De Cooman B, Chen L, Kim H, Estrin Y, Kim S, Voswinckel H (2009) State-of-the-science of high manganese TWIP steels for automotive applications. In: Microstructure and texture in steels. New York, Springer
137. Bian X, Yuan F, Wu X (2017) Correlation between strain rate sensitivity and characteristics of Portevin-LeChátelier bands in a twinning-induced plasticity steel. Mater Sci Eng A 696:220–227
138. von Appen J, Dronskowski R (2011) Carbon-induced ordering in manganese-rich austenite—a density-functional total-energy and chemical-bonding study. Steel Res Int 82:101–107
139. Song W, Bogdanovski D, Yildiz A, Houston J, Dronskowski R, Bleck W (2018) On the Mn-C short-range ordering in a high strength high-ductility steel: small angle neutron scattering and *ab initio* investigation. Metals 8:44
140. Gerold V, Karnthaler H (1989) On the origin of planar slip in fcc alloys. Acta Metall 37:2177–2183
141. Gutierrez-Urrutia I, Raabe D (2013) Influence of Al content and precipitation state on the mechanical behavior of austenitic high-Mn low-density steels. Scr Mater 68:343–347
142. Kandarpa V, Spretnak J (1969) Internal friction from stress-induced ordering of carbon atoms in austenitic manganese steels. Trans Metall Soc AIME 245:1439–1442
143. Rosalie J, Somekawa H, Singh A, Mukai T (2013) Effect of precipitation on strength and ductility in a Mg-Zn-Y alloy. J Alloys Compd 550:114–123

Chapter 8
Conclusions and Final Remark

The book provides an in-depth understanding of nanostructures and nano-engineering approaches in high strength steels. The state-of-the-art of nanostructures is thoroughly discussed, for instance, nanostructures in metallic materials (ultrafine-grained structure, nano-laminates, nano-plates, nano-particles, nano-precipitates, nano-twins, and gradient nanostructure), the technological processes (thermomechanical treatment, solid reaction, severe plastic deformation), the novel nanostructure characterization methods, the fascinating functional and structural properties, as well as its applications. A summary of the state-of-the-art knowledge of the nano-engineering approaches, e.g. precipitation engineering approach, interface engineering approach, short-range ordering engineering approach, was provided. New methods that aid controlling the process of phase transformation during deformation and/or thermal treatment in the steels were focused to be explored. Based on own-related work, the nano-engineering application examples are extended to the transformation-induced plasticity (TRIP)/twinning-induced plasticity (TWIP)/microband-induced plasticity (MBIP) high-Mn steels, medium-Mn steels, Cr stainless steels, Al-alloyed high strength lightweight steels, bearing steels, high entropy alloys, hot work tool steels with nano-sized NiAl intermetallics, Nb/V/Mo microalloyed high strength steels, etc.

The main conclusions are drawn as follows:

- High-Mn austenitic steels are able to combine the excellent formability of an initial fcc microstructure with the extraordinary strain hardening potential of the TWIP mechanism, showing how engineering on a nanometre-scale can improve the plastic deformation behaviour of steels. The main factor which contributes to controlling the strain hardening behaviours of high-Mn steels is the stacking fault energy (SFE). The composition-dependent SFE maps calculated via the subregular thermodynamic model can facilitate in 2D and 3D regimes to predict the deformation mechanisms in high-Mn steels in Fe-Mn-C and Fe-Mn-Al-C systems. The thermodynamic based prediction describes the general features of

W. Song, *Nano-Engineering of High Strength Steels*,
Topics in Mining, Metallurgy and Materials Engineering,
https://doi.org/10.1007/978-3-031-42967-5_8

the strain hardening behaviours, i.e. the transition of TRIP-TWIP mechanisms, the temperature dependency of SFE, the effects of alloy composition on SFE, etc.

- By means of a combined method of ab initio calculations, in-situ high-energy synchrotron X-ray diffraction (SYXRD) and small angle neutron scattering (SANS), the interstitials-contained short-range ordering (SRO) is for the first time observed. Surprisingly, the formation of the highly dispersed SRO in Fe-Mn-Al-C steels enables an increase in strength and ductility either individually or simultaneously. The ab initio (density functional theory) calculations further provide, from a thermodynamical point of view, the assessment of the thermal stability and the ordering of kappa phase formation. Ab initio calculation results show the kappa phase with an ordered structure is more thermodynamically stable than that with a disordered structure. Ab initio calculations reveal that Al has a stronger impact on the ordering of kappa phase than carbon. The long-range ordered kappa phase already starts precipitating from the austenitic matrix as early as 15 min during aging at 600 °C in Fe-30Mn-8Al-1.2C steel. Up to 9 h aging, the lattice misfit between the kappa phase and the austenite matrix still maintains being very small (less than 2%) which may lead to an effective coherent precipitation hardening. The Fe-30Mn-8Al-1.2C MBIP steel exhibits an improved combination of strength and ductility, with respect to conventional high strength low alloy (HSLA), duplex phase (DP) and TRIP steels. The strengthening potential of kappa phase is between 100 MPa to 350 MPa for a volume fraction of approximately 0.05–0.35.
- In the microalloyed 6Mn medium-Mn steels, the Nb or Nb-Mo addition decreases the volume fraction of retained austenite but increases its mechanical stability. The Mo addition to the Nb-bearing steels can lead to a higher density and smaller mean size of the precipitates for the decreased the interfacial energy and enhanced the precipitation kinetic conditions. The Nb-Mo bearing steels present the equiaxed and lamellar mixed microstructure, and exhibit exceptional mechanical properties, i.e. the average products of ultimate tensile strength and total elongation values reach 61,900 MPa%. The increase in yield strength can reach up to 300 MPa by grain refinement and precipitation hardening with the Nb-Mo addition.
- Hydrogen embrittlement (HE) resistance is an important property for modern high strength steels applied in the automobile, energy and chemistry industry. In a cold-rolled Fe-12Mn-3Al-0.05C medium-Mn steel, the combination of austenitization (AUS) and austenite-reversed transformation (ART) produced comparable mechanical properties (ultimate tensile strength (UTS) = 891 MPa, yield strength (YS) = 701 MPa, total elongation = 30.1%) as that in a routine where the ART annealing was applied immediately after cold rolling. The ultrafine-grained martensite colonies provided a large number of interfaces (prior austenite boundaries and lath boundaries) for hydrogen trapping, which increased the hydrogen ingression. ART specimen revealed a clear ductile–brittle transition with increasing hydrogen concentration. Hydrogen embrittlement is considered to be predominated by concurrent contribution of hydrogen enhanced decohesion (HEDE) and hydrogen enhanced localized plasticity (HELP) mechanisms. The AUS + ART specimen exhibited extremely high hydrogen susceptibility of the

ductility regardless of hydrogen concentration. The brittle failure in H-charged samples was attributed to the HEDE mechanism in the (ultrafine-grained) UFG microstructure with a large number of interfaces, and to possible contribution by the adsorption induced dislocation emission (AIDE) mechanism. Consideration of thermodynamic factors suggest that the failure discrepancy in the UFG and non-UFG specimens was likely to be related to the facilitation of hydrogen-rich phase precipitation by interfacial defects.

- SRO strengthening also applies in multi-component alloy systems like high entropy alloys (HEA). At the as-forged state, the single phase CoCrFeNi HEA achieves an outstanding yield strength ($R_{p0.2}$) of 710 MPa at room temperature and 660 MPa at 500 °C. The advanced characterization tools, SYXRD and SANS, are considered to be very helpful to study the lattice distortion and ordering effect, thanking to its high-resolution detection characteristics at nano-/atomic scale and the statistical measurement in large volume in mm^3. It is in particular worthy to be noticed the interesting observation of local ordering effect in the HEAs. Since the local ordering might strongly increase the yield strength and the ductility of the alloys, yet the mechanism is still unclear, it is of high importance to gain further knowledge of how to control and apply this phenomenon in order to achieve enhanced mechanical properties in HEAs in different systems and/or at different states.
- Intermetallic phases have been investigated since long for their behaviour in steels. For a long time, they have been considered critical because of their limited formability and their predisposition for early crack initiation. More recently, new ideas for a beneficial use have come up mainly because of a better understanding of the interaction between microstructure, intermetallic phase formation and mechanical behaviour. Intermetallic phases can be used for further development of conventional steels or the design of new steel classes. The properties are strongly dependent on the alloy content, the matrix crystallographic structure, and the microstructure. Attractive property combinations can be obtained even if the processing route is not significantly changed for cold formable steels both with a ferritic and an austenitic matrix.
- In-situ annealing experiments using synchrotron diffraction depict the separation of {100}-superlattice peaks ascribed to C-enrichment of pre-existing, C-lean $L'1_2$-ordered structures into C-rich kappa phases of alike crystal structure during heating of a high-Al lightweight steel. The peak separation is the first experimental in-situ investigation of the transformation sequence during Fe_3AlC-phase formation. In-situ synchrotron diffraction facilitates the identification of the kappa phase formation sequence, wherein the peak separation of superlattice reflections corresponding to a $L'1_2$-crystal structure was linked to the development of a C-rich kappa phase from an ordered C-lean κ-precursor. The observed transformation sequence suggests a prior ordering of the disordered fcc-phase into a long-range ordered, C-lean κ-precursor, which subsequently undergoes phase separation into a disordered fcc- and C-rich kappa phase.
- With regards to the impacts of tensile deformation levels (15%, 25%, 35% and 45%) on Mn-C SRO, serrated flow and Portevin-Le Chatelier (PLC) bands in

Fe-18Mn-0.6C steel, SANS shows that with the increasing strain during uniaxial tensile tests, the mean radius of Mn-C SRO decreases while the number density of Mn-C SRO increases. The Mn-C SRO in the location after the PLC band passing by exhibits a smaller SRO size with a higher number density compared with that in the section before the PLC band passing by. At large applied strain (>15%), PLC bands form steadily and continuously. On the flow curve, each segment (plateau) between two close serration peaks corresponds to the nucleation and propagation of one PLC band within gauge length. The serration corresponds to the initiation/disappearance of the PLC band outside the gauge length. With increasing strain in the uniaxial tensile tests, the velocity of PLC bands decreases. The longer recrystallization annealing time leads to a higher PLC band velocity. The higher strain rate imposed in the tensile tests, the higher PLC band velocity occurs. The strain distributes heterogeneously and is localized within the PLC bands during plastic deformation. In the section after the PLC band passed by, the local strain is larger than that in the section ahead the PLC band.

To sum up, the new material design concepts at nanometre scale speed up the development of novel metallic materials, mainly steels, with extraordinary properties. These concepts offer new chances for the design of materials by diffusion and partitioning phenomena of interstitial and substitutional elements as well as by interaction of dislocations with a great variety of interfaces. Interface engineering provides great opportunities to enhance mechanical properties by structural refinement via introducing multiple interfaces, i.e. phase boundaries via deformation induced austenite to martensite transformation and nano-twin boundaries via deformations induced twins. Segregation engineering belonging to interface engineering is an effective approach to accelerate the phase transformation via targeted segregation at nanometre scale to defects in nanostructured metals.

The recent boost of the high-tech development and the access to large research facilities with synchrotron and neutron radiation source enable us to process and characterize the metallic materials at atomic scale and in a statistic manner. In order to fully understand the phase transformation and the strengthening mechanisms in nanostructured metallic materials, a combined multi-scale characterization approach is necessary. High-resolution transmission electron microscopy (TEM) enables the study of the microstructural characteristics at atomic scales, such as atoms ordering, defects, and interfaces with detailed crystallographic information. Atom probe tomography (APT) offers the nearly atom-by-atom 3D reconstruction of the material structure for a volume of about 10^6 nm. It provides the great opportunity for the precise chemical composition analysis in specimens in a local chemical atmosphere, particularly the elemental partitioning features at specific locations, such as grain boundaries, etc. SYXRD and SANS are very helpful and powerful tools to gain an overview of the nanostructure in a larger volume. It supports to bridge the characterization and understanding of the materials from atomic scale, up to nano-, micro- and further to macro-scale. In order to study the structure–property relationship in metallic materials, the scale-bridging experiments are highly important.

Nevertheless, there remain still some limitations and challenges for nano-engineering of metallic materials, e.g. quantitative characterization of nanostructures is still one of the most challenging topics and higher resolution of the techniques is required; the understanding of nanostructures in multi-component alloy system is limited, due to the complexity of the system; the scale bridging between atomic level and component level requires fundamental and theoretical description of nanostructures, which to some extent remains blurred in the conventional theory; it is still not feasible to produce metallic materials containing nanostructures in industrial large scales with the existing manufacturing technologies; the knowledge of functional properties of nanostructured materials is limited.

All in all, it is of great interest to further explore the potential of nano-engineering approach in order to develop novel materials to face future global challenges. For example, to develop next generation of GPa steel grades with nanostructures for sustainable development of human society, in particular for automotive, aerospace, infrastructure and energy applications; to combine the nano-engineering approach in an interdisciplinary regime with intelligent production, green-house processing, smart functions, bio-medicare, etc.

GPSR Compliance
The European Union's (EU) General Product Safety Regulation (GPSR) is a set of rules that requires consumer products to be safe and our obligations to ensure this.

If you have any concerns about our products, you can contact us on

ProductSafety@springernature.com

In case Publisher is established outside the EU, the EU authorized representative is:

Springer Nature Customer Service Center GmbH
Europaplatz 3
69115 Heidelberg, Germany

www.ingramcontent.com/pod-product-compliance
Lightning Source LLC
LaVergne TN
LVHW022124191026
850654LV00003B/40
9783031429668